ROAD & TRACK
ON
MAZDA RX-7
& MX-5 MIATA
1986-1991

Reprinted From
Road & Track Magazine

ISBN 1 85520 1186

Published By
Brooklands Books with permission of Road & Track
Printed in Hong Kong

Titles in this series:

Road & Track on Alfa Romeo 1949-1963
Road & Track on Alfa Romeo 1964-1970
Road & Track on Alfa Romeo 1971-1976
Road & Track on Alfa Romeo 1977-1989
Road & Track on Aston Martin 1962-1984
Road & Track on Audi & Auto Union 1952-1980
Road & Track on Audi & Auto Union 1980-1986
Road & Track on Austin Healey 1953-1970
Road & Track on BMW Cars 1966-1974
Road & Track on BMW Cars 1975-1978
Road & Track on BMW Cars 1979-1983
Road & Track on Cobra, Shelby & Ford GT40 1962-1983
Road & Track on Corvette 1953-1967
Road & Track on Corvette 1968-1982
Road & Track on Corvette 1982-1986
Road & Track on Corvette 1986-1990
Road & Track on Datsun Z 1970-1983
Road & Track on Ferrari 1950-1968
Road & Track on Ferrari 1968-1974
Road & Track on Ferrari 1975-1981
Road & Track on Ferrari 1981-1984
Road & Track on Fiat Sports Cars 1968-1987
Road & Track on Jaguar 1950-1960
Road & Track on Jaguar 1961-1968
Road & Track on Jaguar 1968-1974
Road & Track on Jaguar 1974-1982
Road & Track on Jaguar 1983-1989
Road & Track on Lamborghini 1964-1986
Road & Track on Lotus 1972-1983
Road & Track on Maserati 1952-1974
Road & Track on Maserati 1975-1983
Road & Track on Mazda RX-7 1978-1986
Road & Track on Mazda RX7 & MX5 Miata 1986-1991
Road & Track on Mercedes 1952-1962
Road & Track on Mercedes 1963-1970
Road & Track on Mercedes 1971-1979
Road & Track on Mercedes 1980-1987
Road & Track on MG Sports Cars 1949-1961
Road & Track on MG Sports Cars 1962-1980
Road & Track on Mustang 1964-1977
Road & Track on Nissan 300ZX & Turbo 1984-1989
Road & Track on Peugeot 1955-1986
Road & Track on Pontiac 1960-1983
Road & Track on Porsche 1951-1967
Road & Track on Porsche 1968-1971
Road & Track on Porsche 1972-1975
Road & Track on Porsche 1975-1978
Road & Track on Porsche 1979-1982
Road & Track on Porsche 1982-1985
Road & Track on Porsche 1985-1988
Road & Track on Rolls Royce & Bentley 1950-1965
Road & Track on Rolls Royce & Bentley 1966-1984
Road & Track on SAAB 1955-1985
Road & Track on Toyota Sports & GT Cars 1966-1986
Road & Track on Triumph Sports Cars 1953-1967
Road & Track on Triumph Sports Cars 1967-1974
Road & Track on Triumph Sports Cars 1974-1982
Road & Track on Volkswagen 1951-1968
Road & Track on Volkswagen 1968-1978
Road & Track on Volkswagen 1978-1985
Road & Track on Volvo 1957-1974
Road & Track on Volvo 1975-1985

Road & Track – Henry Manney At Large & Abroad

Distributed By

Road & Track
1499 Monrovia,
Newport Beach,
California 92663, U.S.A.

Brooklands Books
PO Box 146,
Cobham, Surrey KT11 1LG,
England

Printed in Hong Kong

R&T

Contents

We are frequently asked for copies of out of print Road Tests and other articles that have appeared in Road & Track. To satisfy this need we are producing a series of books that will include, as nearly as possible, all the important information on one make or subject for a given period.

It is our hope that these collections of articles will give an overview that will be of value to historians, restorers and potential buyers, as well as to present owners of these automobiles.

TESTS: FERRARI 12-CYLINDER 308, BITTER SC 3.9

ALL-NEW MAZDA RX-7, TURBOCHARGED MERCEDES-BENZ 190E

ROAD & TRACK

NOVEMBER 1985 UK £1.60 $2.95

MAZDA'S FABULOUS NEW RX-7

Porsche 944 Performance at an Affordable Price?

27100 06416 11

MAZDA RX-7 GXL

Timid styling, but engineering that raises the standards for sports-car performance

BY RON WAKEFIELD
PHOTOS BY JEFFREY R. ZWART

A CAR WITH a whole bundle of innovative engineering features naturally raises one's expectations when it comes time to get behind the wheel. For the new RX-7, expectations tended to be elevated further even as Mazda was known to be developing it to emulate a much-liked, prestigious and highly capable German sports car.

Said target car was the Porsche 944, which went on the market a year into the RX-7 program. So it should come as no surprise that the new Mazda is precisely the same length as the Porsche, has a nominal 3-bhp addition under its hood, offers token 2+2 seating, bears visual resemblance to both the 944 and big brother 928 and will be offered in a turbocharged version in a few months.

That's not to say the RX-7 is a copy of the 944; far from it, as Jack Yamaguchi's technical analysis documents. On the contrary, Mazda found its own engineering solutions, from the further-developed Wankel rotary engine through the highly original rear suspension and power assisted steering to the automatic adjusting shock absorbers. True to Japanese-industry custom, there's a soup-bowl full of alphabetic names for these features: the engine with its Dynamic Effect Intake (DEI), Electronic Gasoline Injection (EGI) and 6-Port Induction (6PI), Dynamic Tracking Suspension System (DTSS) and the automatic adjusting shocks (AAS), to name just a few.

To find out if the new RX-7 lives up to its many promises, we journeyed to Seattle, where Mazda put a whole flotilla of the new machines at our disposal. In a day of driving over a widely varied circuit of city streets, freeways, secondary highways and winding mountain roads, we tried all the combinations, from a plain-Jane model with unassisted steering and standard tires through automatic- and manual-transmission GXLs to the all-singing, all-dancing, not-yet-available RX-7 Turbo. (For the Canadian market, the base version is plainer and a middle GX model is available.)

Before moving on to the GXL that is the subject of this full road test, then, let's highlight the versions less likely to capture the enthusiast's heart. Until the Turbo appears, every RX-7 is powered by the same 146-bhp 13B rotary engine. The one powertrain variation at the moment is the JATCO 4-speed automatic transmission. We found the automatic to shift smoothly, and its torque-converter lock-up clutch (which operates only in the overdrive 4th gear above 44 mph) isn't as busy as some of today's worst examples of the device. We didn't measure acceleration with the automatic; it probably gives away a couple of seconds on 0–60 time, but the shiftless RX-7 remains quite lively and turns in the same EPA mileages as with the 5-speed.

A few laps of the Seattle International Raceway in the plain-vanilla version with its standard 185/70HR-14 tires on 5½-in. rims revealed predictable, stable handling but squealing rubber. Those interested in extracting the handling potential of this car are therefore advised to order ei-

AT A GLANCE	Mazda RX-7 GXL	Nissan 300ZX Turbo	Porsche 944
Price, base/ as tested	est $17,800 est $19,000	$16,399 $16,399	$22,950 $24,945
Curb weight, lb	est 2700	2990	2790
Engine/drive	2-rotor Wankel/ rwd	V-6/rwd	inline-4/rwd
Transmission	5-sp M	5-sp M	5-sp M
0–60 mph, sec	7.9	8.2	8.3
Standing ¼ mi, sec @ mph	16.1 @ 87.5	16.4 @ 84.0	16.3 @ 84.5
Stopping distance from 60 mph, ft	see text	148	149
Lateral acceleration, g	0.88[1]	0.82	0.82
Slalom speed, mph	62.0	62.3	60.9
Fuel economy, mpg	est 19.0	18.0[2]	22.0[2]

	Pro	Con
RX-7 GXL:	excellent, tossable handling; responsive engine; a sports car of few compromises	body lacks individuality:embodies elements of several other cars; interior materials no better than average quality
300ZX Turbo: tested 1-84	very good V-6 engine, equally good brakes, comfortable ride quality	imprecise handling, severe lift-throttle oversteer, weight, busy interior and exterior design
944: tested 1983 S&GT[3]	excellent, tossable handling; good road feel, superior materials and workmanship	high price, jiggly ride on uneven surfaces, 4-cylinder engine sometimes rough

[1]Mazda test result; [2]trip fuel economy; [3]Road & Track's Guide to Sports & GT Cars

ther the upmarket GXL, which comes with 205/60VR-15s on 6-in. rims, or the optional Sports Package, which is shod the same but includes firmer springs and shocks and an Aero Package. A similar conclusion goes for the nonpowered steering, which is a great improvement over the old RX-7's slack recirculating-ball mechanism and is both adequately quick and manageable in parking; just the same, the exotic new power assisted steering is so good you just gotta order it. Again, it's standard on the GXL but optional on the base car.

So now to our road test car, a GXL with the optional +2 seating and leather upholstery. Thanks to wide doors that reach a bit into the roof, entry into this low sports car is relatively easy, and once inside you find more space than in the old RX-7: an extra inch of head room and leg room as well as 1.5 in. more shoulder room. Because there's also a greater range of seat adjustment, Mazda claims that the new car will accommodate not only taller but also smaller people. For some—primarily those with small children and pets, we think—the available occasional rear seating will add practicality; their backrest folds for maximum cargo space. In the GXL 2-seater, there are locking under-floor compartments.

The standard interior combines vinyl dash and wraparound door bolsters with "velour A" seats and door panels; both materials are unexceptional in appearance but perhaps quite reasonable for the projected less than $12,000 base price. Move up to the GXL and you find a higher-grade velour, more extensive use of carpeting and larger door armrests; still, we found the grains and

textures less impressive and expensive-looking than those, say, of the 944—and were disappointed to find fake stitching all over the place. Our test car had the optional leather, which combines Maserati Biturbo-like crinkled door panels with attractively contoured and stitched seats; not an inexpensive option, but for an elegant RX-7 a must.

In the base model the front seats have only longitudinal and backrest adjustments. The GXL adds adjustable lumbar support and the possibility of raising or lowering the cushions' front end, plus about 2 in. of vertical steering-wheel adjustment. Power seats aren't available. Whatever the version, we found sufficient lateral support in the winged backrests for spirited driving and generally satisfactory seating comfort up front. Because it tilts up and moves rearward on top of the roof, the electric sunroof preserves head room well; the marketing folk haven't yet decided whether to make it standard or optional in the GXL.

Resisting the fad of digitals and graphics, Mazda designers provided a legible, handsome set of instruments. A big 8000-rpm tachometer dominates the instrument cluster; the somewhat smaller 150-mph speedometer is to its right, four quarter-circle gauges to its left—all nicely visible through the steering wheel partly because its center is offset slightly downward. Climate and audio controls are conventionally arranged in the center dash area, where there's also a set of very large air outlets.

Some of the controls are unusual: L-shape levers grow out of the instrument panel for directional signals and high-low beams (left)

and cruise control (right, on GXL). The electric window buttons (GXL) require lifting up, logically, for raising the windows. Other controls—lighting, rear-window defroster, wiper/washer, rear wiper and hazard flasher—are on the instrument nacelle's sides. We found no problem getting acquainted with any of them. Vision out of the RX-7 is good, with the reservation that it isn't easy to judge where the car's corners are for parking. The least satisfactory aspect of this cockpit, in fact, is that the warning lights are out at dash center, out of the driver's direct line of sight.

The new RX-7 may look different, but one thing about it is thoroughly familiar: the unique sound of its rotary engine. Our sound-meter readings don't show it dramatically, but the engine seems much quieter, attesting to the engineers' many sound-reducing measures. It is also a bit smoother and freer-revving, although its rev limit remains at 7000 rpm and a buzzer sounds if you exceed it. On the freeway in 5th gear, engine noise drops to a distant hum, and both wind and road noise are low.

Because curb weight is up (125 lb for the base car, or 5 percent) and torque up 6 percent with little change in gearing, you wouldn't expect much difference from the previous RX-7 GSL-SE's performance feel in normal driving. There isn't much; the engine pulls well from revs that *seem* low revs but aren't (about 2500 rpm), but for real liveliness it's best to make good use of the 5-speed gearbox. This has its rewards; shifting is pleasant, if not the best in the business, and internal improvements not only shorten the shift throws slightly but promise longer 2nd-gear synchronizer life as well. In normal road driving, the only unpleasant characteristic we discovered was a certain drivetrain jerking when getting onto or off of the throttle—something we did not expect given the shock-insulating tube-in-tube (TIT) construction of the driveshaft.

Up through the gears, there's a more noticeable difference as the additional 11 bhp make themselves felt. Our test car reached 60 mph in an impressive 7.9 seconds from rest, 0.6 sec quicker than the old GSL-SE, and covered the standing-start quarter mile in a brief 16.1 sec. Bear in mind that the unadorned $12,000 standard model performs exactly the same; this is a lot of performance for the money. And though our time with the car didn't allow reliable measurement of fuel economy, the new engine boosts EPA ratings by 1 mpg and justifies an estimate of 19 mpg for all-around driving. That should give a reasonable, though not outstanding, range on the 16.6-gal. fuel tank.

Now we come to the most interesting chapter of the RX-7 story, its chassis. In our first miles on public roads in the GXL, we were duly impressed by its solidly constructed body, supple suspension (only slightly less supple in the AAS's Sport setting) and excellent steering feel. Gone is the on-center free play of the old RX-7's steering, but that's only the beginning. The power assist is generous for parking, but falls off smoothly as speed picks up. At freeway speeds there's a reassuring resistance to movements that contributes to a feeling of excellent straight-line stability created by the suspension's utter lack of bump-steer. Unlike its rather noisy, hard-riding and nervous predecessor, the new RX-7 is an ideal car for a fatigue-free cross-country journey.

With the GXL's (fairly) wide tires and wheels and the automatic shocks, it's also an ideal car for the winding mountain road or test laps on the race track. Small steering inputs are rewarded with a crisp, nimble response that Mazda people consider very important to sports car drivers (we agree); and yet true to its promise, the new suspension delivers safe, reassuring basic understeer when the "g's" build up. If ever there was a forgiving car, this is it. You can corner at speeds higher than you'd expect with only 205-mm tires, without squeal or drama. And if you try the old expert-driver tricks—"tossing" the car into turns to set up some initial oversteer, or lifting off the throttle at mid-turn to get the tail out—you find that they work, but just enough. Nothing unduly sharp or at all surprising ever happens; the rear end just comes out smoothly, as much as you want it to and not a bit more. We found steering feel wholly natural under these conditions too, which is almost surprising in view of all the technological tricks that went into it.

To the extent that we could get them, the numbers look good too. Although we were not able to put the GXL through our normal skidpad test, Mazda claims 0.88g. We saw this reading on our accelerometer at SIR and think it may be conservative in skidpad terms, for the rather crudely suspended GSL-SE predecessor turned in a very good 0.86. In the slalom test, the Mazda also acquitted itself very well if not sensationally, weaving through the cones at an even 62.0 mph; the benchmark 944 did 60.9 mph, although quite a few cars can top 62 mph these days.

Unfortunately, we were not able to find out if the GXL's impressive-looking brakes deliver on their promise. Our test car had a hard day of track driving by a number of journalists (including ourselves) behind it, and as soon as we began our brake tests it was apparent that its brakes were pretty well finished. There was no time for repairs, so for now we'll have to confine ourselves to impressions gathered in that day of 10/10ths driving at SIR. Under those hard conditions, we never experienced any fade or uncertainty; the four vented discs hauled the fast-moving GXL from speeds in the 80–110-mph range to tight-corner speeds repeatedly, and in our own fade tests—which we *did* successfully complete—they also failed to show any fade. It is too bad, though, that no anti-lock braking system is available right now; Mazda promises to have one for the 1987 model year.

Besides driving the new RX-7, we also tried its principal competitors, the Nissan 300ZX and Porsche 944 (yes, we know the 944's price is much higher and we'll come to that). Without jumping to sweeping conclusions that only a proper comparison test can deliver with certainty, we can make a few observations:

Despite the 300ZX's brilliant engine, the new RX-7 musters comparable performance; and in the chassis department there's no comparison. Where the RX-7 tracks straight no matter what the surface, the Nissan tends to wander from bump to bump, requiring a lot of attention from its driver. On the track, though it handles fairly well, the Nissan suffers a classic case of the off-throttle oversteer common to semi-trailing-arm rear suspension.

And what of the 944, which served as Mazda's *Vorbild* but costs about $3000 more than the GXL's (estimated at this writing) price with air conditioning? Well, there's no denying that it's finished better than the Mazda, though the pilot-production RX-7s we tested probably weren't fully up to the quality level you can expect from production cars; in particular, its interior—though just as plastic as the RX-7's—looks far richer. But when it comes to performance and handling, the new RX-7 has at least a slight edge. Its 0–60 and quarter-mile times are better; its engine notably smoother. Though we don't have firm skidpad figures, we did drive both at SIR and on the road and found the Mazda's handling and ride more refined. (A racing driver present timed his laps on the Washington raceway and found an RX-7 with the optional Sports Package slightly faster than the 944.)

Another car we drove was the upcoming RX-7 Turbo, though not necessarily in its final production form. At SIR, the turbo engine made much higher speeds possible on the straights and thus worked the brakes a good deal harder; but they seemed up to it. Experimenting with the throttle at lower speeds, we found turbo lag to be minimal. With "only" 182 bhp, the RX-7 Turbo probably won't be as close to its Porsche counterpart (the 944 Turbo) as the GXL is. More on this version when it hits the market.

The very fact that the new RX-7 can be compared with a car like the Porsche 944 with a straight face means that Mazda has made some real sports-car progress here. The styling may not be very original, but thanks to its innovative new engineering, it delivers brilliant performance and refinement for a starting-at-$12,000 sports car. Upgraded to GXL trim and fully equipped, the RX-7 should come in somewhere under $20,000 (excluding dealer gouging) and deliver an impressive lot of performance and luxury at that level. And the previous RX-7's excellent reputation for reliability is not to be ignored.

Our preliminary conclusion about the new RX-7, then, is that Mazda has raised the standards of sports-car performance by a sizable margin.

ROAD TEST

MAZDA RX-7 GXL

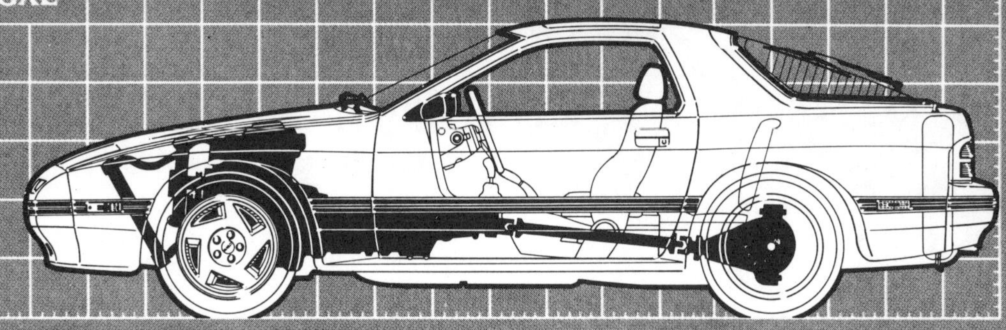

SCALE: 10 in. (254 mm) DIVISIONS
DRAWING BY BILL DOBSON

PRICE

	RX-7 GXL 1986	RX-7 GSL-SE 1985
List price, all POE	est $17,800	$15,095
Price as tested	est $19,000	$16,095

Price as tested includes std equip (air cond, AM/FM stereo/cassette, elect. window lifts, elect. sunroof, auto. adj shocks, rear wiper, elect. mirrors, cruise control), leather front seats (est $700), 2+2 rear seating (est $500)

IMPORTER

Mazda Motors of America (Central) Inc, 3040 E. Ana St, Rancho Dominguez, Calif. 90221

GENERAL

Curb weight, lb	est 2700	2640
Test weight	est 2885	2825
Weight dist (with driver), f/r, %	50/50	54/46
Wheelbase, in.	95.7	95.3
Track, f/r	57.1/56.7	55.9/55.1
Length	168.9	170.1
Width	66.5	65.7
Height	49.8	49.6
Ground clearance	5.8	6.1
Overhang, f/r	34.5/38.7	35.7/37.7
Trunk space, cu ft	11.3 + 7.8	23.4
Fuel capacity, U.S. gal	16.6[1]	

ACCOMMODATION

Seating capacity, persons	2+2	2
Head room, f/r, in.	36.5/32.0	35.5
Seat width, f/r	2 x 20.0/2 x 17.0	2 x 20.5
Seatback adjustment, deg	40	55

ENGINE

Type	2-rotor Wankel	
Bore width x chamber major axis, mm	80.0 x 240.0	
Displacement, cc	1308	
Compression ratio	9.4:1	
Bhp @ rpm, SAE net	146 @ 6500	135 @ 6000
Equivalent mph	159	132
Torque @ rpm, lb-ft	138 @ 3500	133 @ 2750
Equivalent mph	85	60
Fuel injection	Mazda electronic	
Fuel requirement	unleaded, 87 pump octane	

Exhaust-emission control equipment: 3-way catalytic converters, exhaust-gas recirc, oxygen sensor

DRIVETRAIN

Transmission .... 5-sp manual

Gear ratios:

5th (gear) overall	(0.71)	2.91:1	(0.76)	3.10:1
4th	(1.00)	4.10:1	(1.00)	4.08:1
3rd	(1.37)	5.62:1	(1.42)	5.79:1
2nd	(2.00)	8.20:1	(2.19)	8.94:1
1st	(3.48)	14.27:1	(3.62)	14.77:1
Final drive ratio		4.10:1		4.08:1

CALCULATED DATA

Lb/bhp (test weight)	19.8	20.9
Bhp/liter	111.6	103.2
Engine revs/mi (5th gear, 60 mph)	2450	2750
R&T steering index	0.87	1.17
Brake swept area, sq in./ton	293	na

[1]Single entry indicates specifications are identical.

CHASSIS & BODY

Layout	front engine/rear drive	
Body frame	unit steel	
Brake system, f/r	10.9-in. vented discs/ 10.7-in. vented discs, vacuum assisted	9.8-in. vented discs/ 10.1-in. vented discs, vacuum assisted
Swept area, sq in.	418	na
Wheels	alloy, 15 x 6JJ	alloy, 14 x 5½JJ
Tires	Bridgestone Potenza RE71, 205/60VR-15	Pirelli P6, 205/60VR-14
Steering type	rack & pinion, electronic variable pwr asst	recirc ball
Overall ratio	15.2:1	variable, 17.0–20.0:1
Turns, lock-to-lock	2.7	3.7
Turning circle, ft	32.2	31.5
Front suspension:	MacPherson struts, A-arms, coil springs, tube shocks, anti-roll bar	MacPherson struts, lower lateral links, compliance struts, coil springs, tube shocks, anti-roll bar
Rear suspension:	trailing arms with camber-control arms on drop links, semi-trailing links, floating hubs, coil springs, tube shocks, anti-roll bar	live axle on lower trailing links, upper angle links & Watt linkage, coil springs, tube shocks, anti-roll bar

ROAD TEST RESULTS

ACCELERATION

	RX-7 GXL 1986	RX-7 GSL-SE 1985
Time to distance, sec:		
0–100 ft	3.2	3.3
0–500 ft	8.7	8.8
0–1320 ft (¼ mi)	16.1	16.4
Speed at end of ¼ mi, mph	87.5	84.0
Time to speed, sec:		
0–30 mph	2.4	2.6
0–60 mph	7.9	8.5
0–100 mph	24.7	26.2

SPEEDS IN GEARS

5th gear (5250/5740 rpm)	128[1]	126
4th (7000)	121	117
3rd (7000)	88	82
2nd (7000)	60	53
1st (7000)	35	33

FUEL ECONOMY

Normal driving, mpg	est 19.0	18.5

HANDLING

Lateral acceleration, 100-ft radius, g	0.88[1]	0.86
Speed thru 700-ft slalom, mph	62.0	61.1

BRAKES

Minimum stopping distances, ft:		
From 60 mph	see text	157
From 80 mph	see text	282
Control in panic stop	very good	
Pedal effort for 0.5g stop, lb	22	27
Fade: percent increase in pedal effort to maintain 0.5g deceleration in 6 stops from 60 mph	nil	
Overall brake rating	very good	

INTERIOR NOISE

Idle in neutral, dBA	53	52
Maximum 1st gear	73	74
Constant 30 mph	67	64
50 mph	71	70
70 mph	74	75

[1]Mazda test result

ACCELERATION

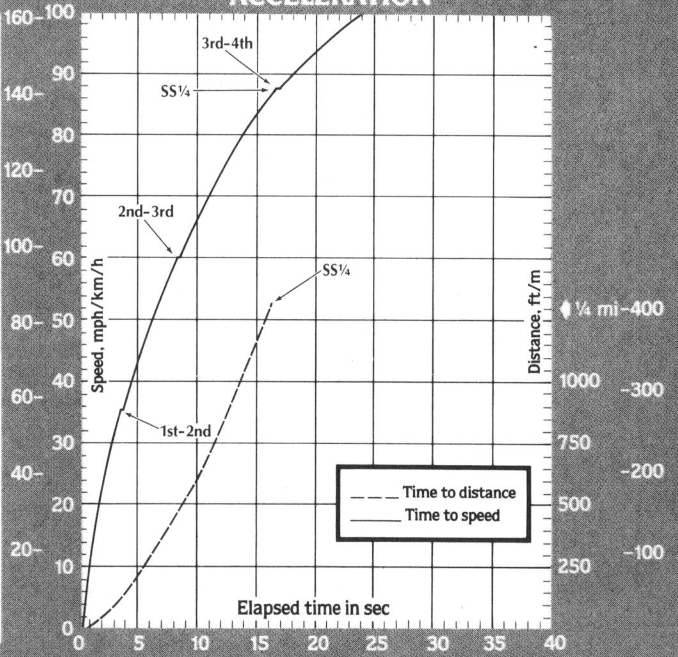

Scoop of Turbo is functional, leading directly to intercooler. Although resembling that of ZX, it is wider, nearer to center of hood, but still offset.

Windshield, A-posts, doors and side glass have Laser/Daytona character, with very aerodynamic results.

Nose is clean, efficient, but lacking in character. Fender flares are horizontal, with crisp upper edges, similar to those of Porsche 944, Mercury Capri.

Styling Analysis

THE 1986 MAZDA RX-7 is more contemporary than the familiar 1978–1985 model it replaces, but at what cost in individuality? The older car, while by no means a standout in styling finesse, had a high-recognition factor that did it a lot of good. Everyone knew exactly what an RX-7 was and, more importantly, that Mazda knew exactly what it wanted the RX-7 to be. Now, that's not as clear.

The trouble with extensive marketing surveys and the slavish incorporation of "must" features is that a lot of excellent cars come out looking like a lot of other excellent cars. If you consider the 1986 RX-7 in three parts—front, middle and rear—you have the ghosts of three not-yet-departed GT coupes occupying the same road space: the nose of the Porsche 944, the doors and roofline of the Chrysler Laser/Dodge Daytona, and the rear deck of the Chevrolet Camaro/Pontiac Firebird. That the Mazda blends these elements successfully is certainly an accomplishment; it's a handsome, well proportioned coupe. But will you recognize one the first time you see it on the road? Will you even pay attention when the tenth one goes by?

In fairness, the RX-7 is considerably smaller than the Camaro/Firebird and a bit smaller than the Laser/Daytona, both fairly roomy 2+2s; so it will look a little more sporting on the road. But it is just about the same package as the 944, even to the latter's pretense at +2 seating, and it is very likely to be mistaken for the German car (especially its fender flares) at certain angles. And a few details, such as the front bumper and the design of the optional 14-in. alloy wheels, owe something to that other Porsche, the 928. The GXL wheel design looks like it came from the Nissan ZX, as does the hood scoop of the Turbo model (to be introduced later). Is this really what Mazda intended? It seems so, and this RX-7 reinforces the depressing trend toward the generic GT. There's not one single styling feature that is uniquely Mazda's, and no obvious continuity with the previous car; only the rear end bears a vague resemblance. Personally I'd have preferred that Mazda take a styling trip further down its own special road.

The two versions of the RX-7 differ mainly in wheel styles, but the GXL also has an additional side molding and a rear wiper. Inside, there are different steering wheels and upholstery, with leather an option on the top model. The businesslike instrument panel and console are not so much styled as well designed. The cockpit will satisfy admirers of the previous RX-7, with its very classic layout.

After all this criticism of the new RX-7's lack of originality, it might seem that I don't like it. In fact, it is a good, clean, basically honest design that I wouldn't be the least embarrassed to be seen in. And I guess my feelings wouldn't be hurt if you didn't know for sure it was an RX-7.—*Jonathan Thompson*

Cast aluminum 16 x 7J wheels of Turbo are the most interesting design, with 10 angled slots near rim.

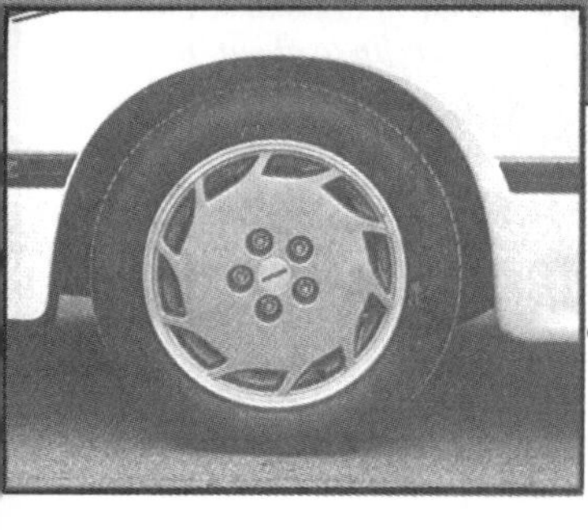

The 14-in. alloy wheels, optional on the standard model, are Porsche-like.

Mazda RX-7 Technical Analysis

Tell someone to think of Mazda and innovation, and even money says they'll say rotary engine. The rotary has been Mazda's calling card since coming to the U.S., and although there have been plenty of pieces in a Mazda other than the engine, none of them has been especially clever.

Until now. For the first time, Mazda has produced a chassis that has all the novelty of the rotary engine, and it is just as baffling to describe. It is also the newest, most fascinating part of the all-new RX-7, so a technical description of the car should begin with the suspension.

Starting at the front, MacPherson struts are used, but with some nice touches added. The wide-base A-arms are H-section aluminum forgings, which cost twice as much as steel arms and save 4.4 lb of precious unsprung weight.

The old RX-7's recirculating-ball steering has been replaced by a rack-and-pinion system. That alone should eliminate the on-center steering imprecision RX-7s are known for, but Mazda also made the center of the steering rack slightly thicker than the ends, so there is no slop in the on-center steering. Special machining allows a strong spring to preload the pinion to the rack. Steering ratio is 20.3:1 when not assisted; this drops to a quick 15.2:1 ratio for the electronically regulated power assisted steering.

Yes, a computer controls the steering assist. The Mazda's steering assist decreases with speed, until there is virtually no assist above 112 mph. The steering computer also receives signals of the steering angle. On the basis of steering angle and vehicle speed, the computer determines the lateral acceleration and regulates the level of assist to keep effort at the steering wheel linear as lateral acceleration increases; normally, effort goes up more than in proportion to "g," but the Mazda engineers feel strongly that this is the best basis for steering feel. But that's not all. The Mazda's steering computer can also sense decreasing surface friction. It also determines (via hydraulic pressure) the force at the road and compares it continuously to a standard data base. When less assist is needed to maintain a given steering angle at a given road speed, the computer decides that there is a lower coefficient of friction. It then increases the amount of pressure assisting the steering, decreasing the effort needed at the steering wheel. This tells the driver that the surface is slippery.

Now if you've mastered that, you may be prepared for an introduction to beginning 4-wheel steering, as seen on the new RX-7. Mazda, you may recall, has experimented with electronically controlled 4-wheel steering. The new RX-7 doesn't get this, but it does have a rear suspension that varies rear-wheel toe angles according to handling requirements.

Nominally, the rear suspension is of the tried-and-true semi-trailing arm type. But there's much more to it than that. The system responds to initial steering inputs with slight toe-out. This quickens steering response, as the back end of the car is turned outside the path of travel—remember that it's the outside tire that takes most of the cornering load, so for small deflections rear toe-out functions as 4-wheel steering. But when lateral acceleration rises above 0.4 to 0.5g, the rear wheels go through neutral toe, and then to increasing amounts of toe-in as lateral forces increase more. This increase in toe-in reduces oversteer at maximum cornering speeds. During sharp acceleration, engine braking, and braking, the rear suspension

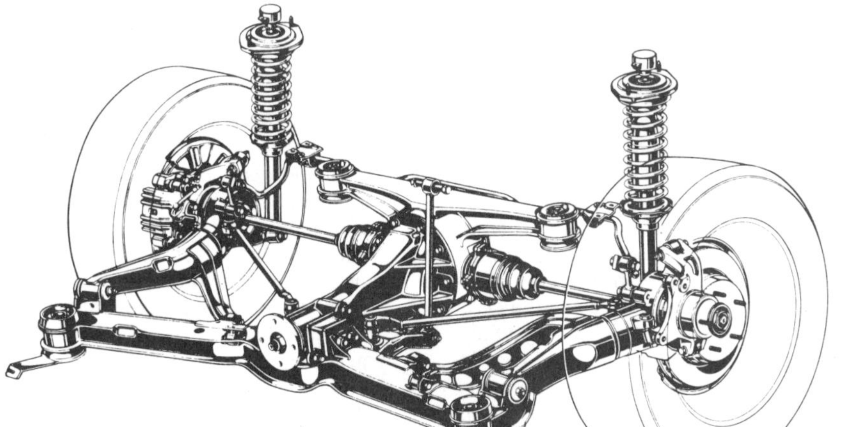

Rear suspension.

also allows toe-in for increased stability.

A unique "triaxial floating" rear hub allows all this carefully controlled monkey motion. The patented Kijima-Maebayashi hub, so-named for its inventors, is attached to the semi-trailing arm at three points. One of the points is a ball-joint that allows movement in any axis. The other two points have rubber bushings that control movement very specifically in different directions. It is the forward bushing that limits toe. It is preloaded to create the initial toe-out; after the preload is overcome, it begins to allow forces on the tire to cause the desired toe changes. Normally, semi-trailing-arm rear suspensions allow disturbing changes in camber, but here too Mazda went a new route. What looks like a conventional semi-trailing arm is actually only part of it, with its inboard side functioning as a camber control link with the help of a small vertical link between it and the underbody. A separate angled lateral link completes the actual semi-trailing arm. Virtually constant camber (-1.5 degree) is claimed to be the result. Mazda calls the whole affair Dynamic Tracking Suspension System.

Working with the sophisticated chassis is something called Automatic Adjusting Suspension (AAS), a computer-controlled system for the shock absorbers on the upscale GLX. Pushbuttons in the car select Normal or Sport modes. In Normal, shock damping is kept at the softest level for normal straight-line driving, except that above 50 mph the front shocks are stiffened to "firm" for better directional stability. In the Sport setting, all four shocks stay on the "firm" setting most of the time. In either setting, all four shocks stiffen to "very firm" any time there is more than a preset level of acceleration, lateral acceleration or braking.

Two braking systems are offered. On the base model, vented front discs and solid rear discs are used with cast-iron single-piston floating calipers. On the GXL all the discs are vented and 4-piston aluminum calipers are used in front. The low-level Mazda also has small 185/70HR-14 tires on 5½-in. wide steel wheels; alloys are optional. The GXL gets 205/60VR-15 tires on 6-in.-wide aluminum wheels.

Of course, the most noticeable change in the new RX-7 is the new body, which was shaped to meet the various demands on the car. For instance, in Japan the tax on a car doubles if car width exceeds a certain figure, so the new RX-7 is only 66.5 in. wide. Aerodynamic requirements added their own styling limitations. Claimed C_X is 0.31 for the standard RX-7, or 0.29 when the car is equipped with an optional 4-piece aerodynamic package. Lift coefficients are 0.08 front and 0.14 rear; the discreet aero package reduces the rear lift coefficient to 0.07.

Toe Control

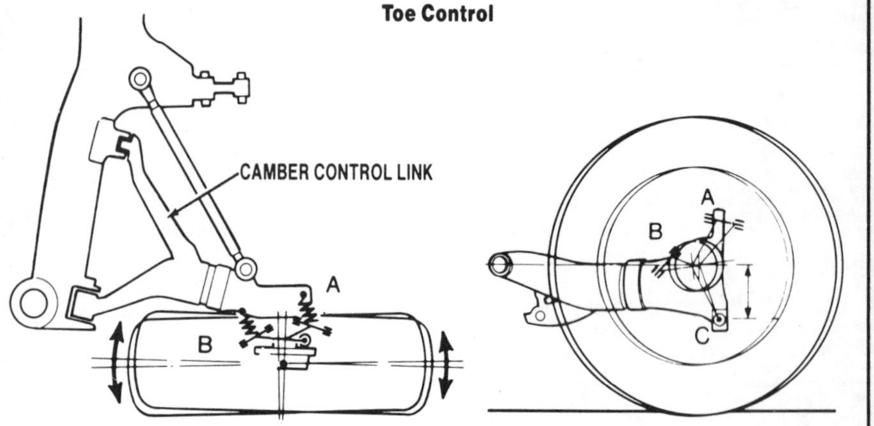

The RX-7's floating hub is joined to the rear suspension by bushings A and B and uniball joint C. Under acceleration, propulsive forces (acting at the wheel center) misalign with the more inboard bushings, causing the wheel to assume a slight toe-in. Braking also produces toe-in: Points A and C define the toe change axis, which is inclined significantly inboard from vertical. If you imagine this axis extended to intersect the tire contact patch, the intersection occurs outboard of the tire's geometric center—where the braking force acts. Hence, under braking, the wheel again pivots to toe-in. Engine braking (another force acting at the wheel hub) causes the floating hub to rotate clockwise around point C, distorting the upper bushings to produce the desired toe-in.

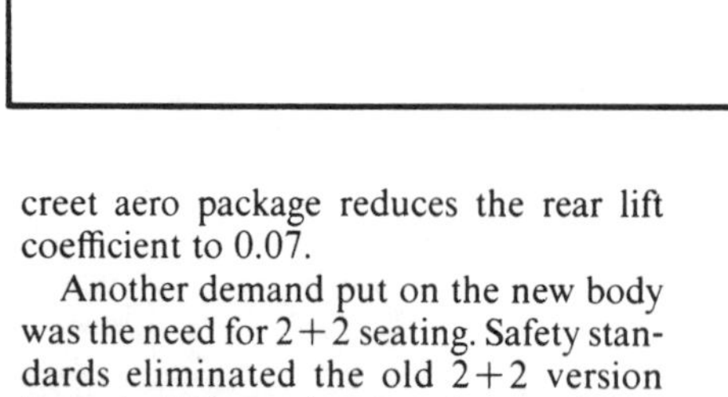

Finally, we come to cornering: Lateral forces on the outside tire naturally compress that suspension's compliance elements, pivoting the entire semi-trailing arm system to produce toe-out at cornering levels up to about 0.2g. Beyond that, pre-loaded bushing B begins to distort, levering the wheel around axis A-C to return it to toe-in. Under lateral load, semi-trailing-arm suspensions without this feature typically develop increasing amounts of toe-out.

Another demand put on the new body was the need for 2+2 seating. Safety standards eliminated the old 2+2 version from the U.S. market, but the new RX-7 now has adequate protection for rear-seat passengers, so optional rear seats are available. But they are strictly occasional—as you'd guess, since the new car is actually an inch shorter than its predecessor.

Rear-seat access requires larger door openings; single-piece door panels are used because they combine light weight and adequate strength. These, Mazda claims, remain fully sealed even at 150 mph in a 22-mph crosswind.

A soft-faced 2.5-mph front bumper is very 944 Turbo-like, but there's a reason for this. During testing in the American desert, Mazda engineers found that flying rocks scratched any metal surface, so the soft facia was developed.

An important part of the new car's development was weight control. Despite being essentially the same size as the previous RX-7, weight has crept up from about 2500 lb to about 2625 lb for the base model. Much of this additional weight comes from the heavier rear sus-

The turbocharged version of the 13B rotary displays a sizable air-to-air intercooler atop the engine.

pension, larger wheels and tires and bigger brakes; but the body structure has larger openings and a heavier single-piece rear glass hatch that all add pounds. The concern with weight was so great that part of the way through the development process, all design work was stopped while the company considered what to do about the car's growing weight. Every engineer involved with the car was asked to find one gram of weight to remove, and as a result, the RX-7 has an exceptional number of lightweight components.

Aluminum alloy is used for front suspension A-arms, engine-mounting brackets, front brake calipers, final-drive housing, spare wheel, hubs and the hood of some models. Even the jack is an aluminum part made in Germany; it saves 3 lb compared with a steel one. Weight distribution, with two people and fuel, is near-ideal at 50.2 percent front and 49.8 rear.

Inside the RX-7 are more new pieces, but they are not drastic revisions. There are still round instrument dials and a center console resembling current RX-7 styling. The seats have additional support and adjustments; leather upholstery is optional in the GXL. The heater and air conditioner have greater capacity and there's more room for people, but despite the improvements in comfort, Mazda has avoided creating a small luxury car.

To preserve head room when a sunroof is fitted, the movable roof panel lifts and slides back above the roof. A theft-deterrent system, optional in the GXL, flashes lights and sounds the horn if someone forces the hood, doors or rear hatch, or tries to hot-wire the ignition. The ignition is also disabled by theft attempts and can only be reset by the key.

Electronic controls aren't limited to suspension and security. The RX-7's rotary engine now benefits from a powerful engine management computer that regulates ignition and fuel injection. There are also mechanical improvements, resulting in better power and economy as well as lower noise.

All models of the new RX-7 are equipped with an updated 13B engine. It is the 13B, you may recall, that powers the top-of-the-line GSL-SE model currently. The smaller displacement 12A engine is now out of the lineup.

Horsepower of the new 1.3-liter engine is 146 bhp at 6500 rpm, up 11 bhp from the current 13B. Torque is 138 lb-ft at 3500 rpm. Like the existing 13B engine, the new RX-7's engine has three intake ports per rotor for best breathing. What Mazda calls Dynamic Effect Intake system (a design that uses the strong intake pulse waves of one rotor to help fill the other rotor) has been revised for more efficient airflow. Dual fuel injectors are now used for each rotor, with the secondary injector only operating at high power. New casting techniques reduce the weight of the rotor 14 percent, while thinner 3-piece apex seals improve sealing and reduce friction. The rotor housing is now finished with a porous chrome surface that retains lubricant for improved sealing.

In the exhaust port of the 13B engine is a multichamber insert designed to control the rotary's sound intensity. The piercing quality of the rotary's engine noise is caused by the rapid opening of the exhaust port as the rotor slides past it, according to Mazda. By diffusing the sound waves, the Mazda's exhaust note is improved.

A turbocharged version of the new 13B engine has been designed, but it won't be offered in the first RX-7s sold in the U.S. Mazda isn't yet saying when the turbo RX-7 will be sold, But some turbo details have been revealed.

Performance of the turbocharged car is expected to be comparable with that of the turbocharged Porsche 944. Acceleration to 60 mph requires 6.7 seconds. The quarter mile is a 15.2-sec event. Top speed, Mazda claims, is well over 150 mph.

The turbo produces 182 bhp at 6500 rpm and 183 lb-ft of torque at 3500 rpm.

Dynamic Effect Intake system

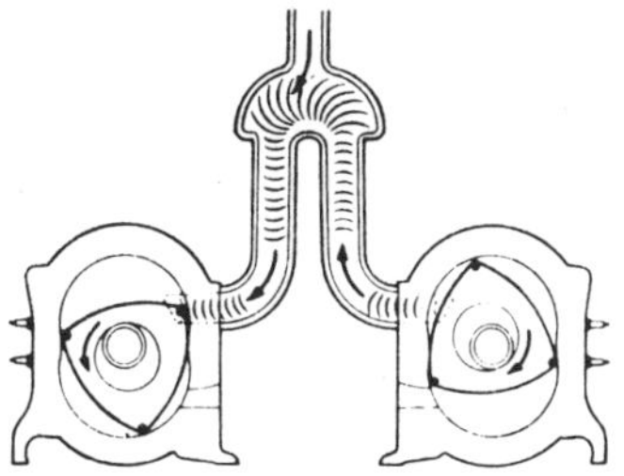

The new 13B engine's intake tract takes advantage of pressure waves created at port closure to improve chamber filling. Nuances in the design extend this resonance "supercharging" to the widest possible rpm range.

This power level is significantly lower than the Porsche 944 Turbo's 220 bhp and 243 lb-ft of torque, but the Mazda is a lighter car with less aerodynamic drag.

Mazda has now gathered a large amount of turbocharger knowledge, and the installation in the RX-7 is particularly thoughtful. Thanks to an innovative twin-scroll layout of the RX-7 turbocharger, the intercooled turbo engine produces more torque at 1500 rpm than the naturally aspirated engine at its torque peak. From 1500 rpm to the redline of 7000 rpm, torque never falls below 145 lb-ft. These are pretty astounding figures.

The exhaust has two ports, a large secondary and a small primary. Intake vacuum keeps the large secondary closed at low rpm, so exhaust enters the turbo at relatively high velocity. When the engine is pumping enough air, the secondary opens. The two exhaust channels find their way into the turbocharger at different angles, so that they hit the turbo's blades at the most effective angles. And on top of all this, a rotary's exhaust is especially effective in spinning a turbocharger because of those high-impact exhaust waves—the ones that require the multichamber insert in the normal engine's exhaust to muffle noise.

All of this had to be done without turning the Mazda into what the EPA would call a Gas Guzzler; because of its particular trauma in the Energy Crisis days, Mazda is sensitive about this in a way no other company can imagine. The normal RX-7 returns EPA figures of 17 city/24 highway, an improvement from the current car's 16/23 figures.

Transmissions are 5-speed manual standard; a 4-speed automatic with centrifugally engaged converter lockup only in 4th gear is optional.

It's clear that Mazda has addressed every shortcoming of the previous RX-7 and has done so with cleverness and skill. If the new RX-7 owner is asked to think of Mazda and innovation, perhaps he will think of a lot more than just the rotary engine.—*Jack Yamaguchi*

MAZDA RX-7 (NEW) VS

PORSCHE 944 (IMPROVED)

Sometimes more is better, and sometimes more is too much

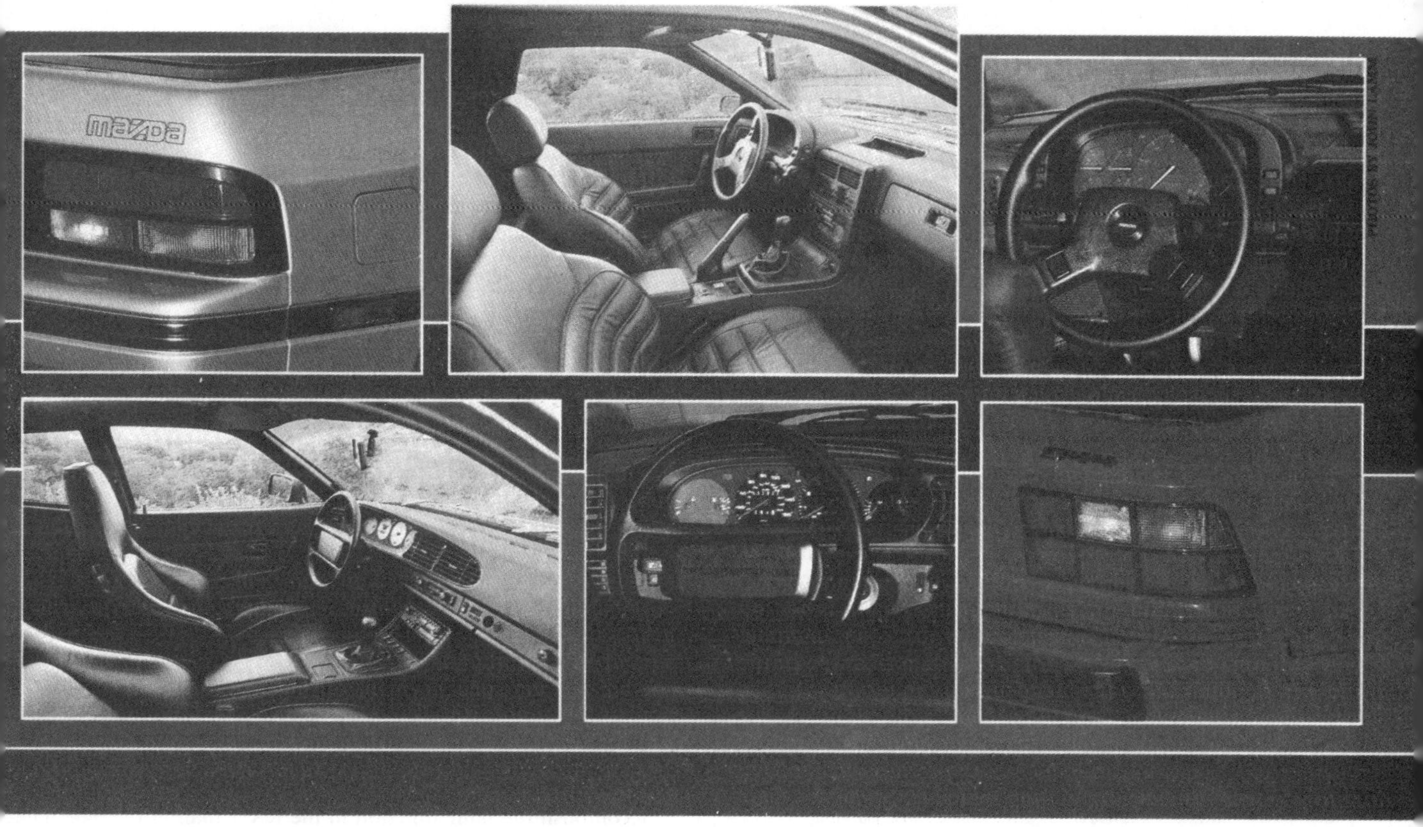

STANDING OUT FROM the crowd is a good way to become both a success and a target. When Porsche transformed the mediocre 924 into the 944, the new car made its buyers happy, put the factory solidly at the top of the $25,000 sports/GT class and gave everybody else something to shoot at.

The biggest gun is aimed by Mazda: the new RX-7. When that car was introduced, Mazda's spokesmen were cheerfully frank. They reckoned they could do the same thing, but better. Their new car isn't an imitation, rather it's the same basic idea but with some different techniques and technology that let the RX-7 equal the 944 while selling for less. Or so they said/hoped/claimed.

It was an open, honest challenge, one with which the Marquess of Queensberry would find no fault. Nor is there a question as to sticker prices.

Car versus car is more complicated.

The Porsche 944 is a textbook example of improvement. Its beginning was the 924, itself Porsche's entry into the less-than-expensive GT market, configured as the Other Kind Of Porsche, with front engine driving the rear wheels, and with water cooling, as best known with the 928.

The 944's 2.5-liter inline-4 is basically half the 928's V-8. The model was given major improvement in 1985, with a rear suspension and mounting reworked better to control the rear wheels and reduce the noise and harshness transmitted from the rear suspension to the car's interior. Said interior at the same time got a smaller steering wheel (radical move, for a German car) and rearranged instrument panel. The seats were given a wider range of adjustment and virtually the complete insides were moved about for more room and convenience within the same overall size.

The RX-7 is new. The configuration shares features with the 944, as in front engine, rear drive, two adult seats in front and (in some models) padded pockets that pose as occasional or kiddy seats in the rear. Both cars have large rear hatches, both come with air conditioning, deluxe radio, sunroof, all the things people like, either as standard, as part of a package or as an option.

The RX-7's exterior measurements are very close to those of the 944. The styling is not quite as close, as the RX-7 looks like a 944 in the front, a Dodge Daytona in the middle and a Chevrolet Camaro in the rear.

None of this happened by chance. Mazda isn't a dominant force either in the world or at home in Japan. The company is growing and the growth is part of a plan. Mazda made its initial foray in the U.S. with quick little coupes and sedans that offered value and the novelty and performance of the Wankel-licensed rotary engine.

When Mazda introduced the RX-7 in 1978, the car was a move up, another part of the plan. That first sports/GT was aimed at the Datsun (now Nissan, of course) Z-car. Datsun had tapped a market for such cars by offering Jaguar- or Corvette-like attributes for less money, so Mazda built one for even less. And it worked.

By now Mazda has a good reputation, for value and for the reliability of its rotary engine. Thus Mazda's next move up, to challenge Porsche instead of Nissan.

Mazda people say the RX-7 is no mere copy of the 944. They're right. The RX-7 is different.

The differences involve three areas.

First is simply the rotary engine, a problem solved by throwing money, brains and time into it. The 13B rotary is almost impossible to break. In the form used for all the RX-7 models, it's tuned and improved and given a few more bhp, 146 for the RX-7 against 147 for the latest 944 engine.

Second is a different rear suspension design. Really different. The RX-7's rear suspension isn't just independent with trailing arms, it has a geometry that steers the rear wheels.

In two directions. When the front wheels are steered, the initial roll of the car that results from that action causes the outside rear wheel to steer opposite to the direction of the front wheels, albeit at a much less acute angle. The front of the car doesn't move over; the car describes an arc.

That's the first stage. But as lateral g levels increase, the rear-wheel motion in effect reverses itself, aiming the rear wheels toward the turn and creating enhanced stability by promoting a little more understeer.

The third difference involves optional power assisted steering and suspension with adjustable shock absorbers.

The original RX-7 had recirculating-ball steering, criticized for lack of feel and precision. The new car gets rack-and-pinion steering. With power assist, it also gets computer control, which regulates the amount of boost according to the speed of the car, the cornering force and the grip the tires have on the road. The feel the driver gets from steering is, in effect, controlled by what the "assis- ➺

tant" is doing, that is, steering effort is light in parking maneuvers because the speed is low, it gets progressively heavier as cornering loads build, but it's light on the edge of traction because it senses the tires haven't much grip left.

In large measure the driver thus drives at one remove: The driver tells the steering wheel what's wanted and the wheel reports to the computer delivering the forces to the valves and servos actually linked to the front wheels.

The adjustable suspension is just as calculated. There are two driver-controlled settings—Normal ("soft") and Sport ("hard")—with a button on the console. And a computer can override either of these settings into "hard" or "very hard" modes as a function of a variety of dynamic conditions. The computer controlling this system is linked to the brakes and accelerator as well as the steering. If the car is going 50 mph or faster, the front shocks are set on "hard" no matter what the switch says. If the thin pedal is fully down and the speed is below 30 mph, all shocks are set "very hard" to reduce squat. If the brakes are applied hard, all shocks go "very hard" to minimize dive. If the car corners hard, they switch to "very hard." How does it know? Easy: by sensing steering wheel angle and road speed, then calculating lateral acceleration.

The strategy was formulated back in the lab and on the factory's proving ground. The intention was to give the RX-7 superior handling and ride under the wide variety of conditions to which the car would be subject, with the (perhaps unwanted) side effect of making the driver less of the equation.

This is a mixed blessing, as we'll see.

In a direct, factual, specific comparison, the RX-7 and 944 are remarkably similar, even allowing for one being used as the rough draft for the other.

Overall length, curb weight, wheelbase, rated power, drivetrain, power-to-weight ratio, weight distribution, etc, are within fractions of each other.

Not surprisingly, the actual performance of the two cars is nearly as close. Acceleration times, stopping distances, top speed and so on are as one would expect from comparing the specifications.

But they are remarkably different when driven, far more than the looks or measurements or acceleration times prepare one for.

Starting from a blank slate, the Porsche feels small and solid. The steering requires effort, so do the clutch pedal and gearshift lever, while the engine needs to get into the middle of its rev range before there's enough power to move off the mark.

On a smooth surface interrupted by seams and dots, the ride is firmer than firm, almost harsh with lots of noise coming from the tires as they thump across markers. The newest 944 is much improved from the original, but it's still obvious that the designers built a sporting chassis and suspension, and if you don't want to accept the limits, perhaps you aren't ready for a Porsche.

The Mazda feels lighter, especially in the steering and other controls. Longer, as pedals and lever both need more travel to accomplish the same function in the RX-7 than in the 944. The Mazda's ride is smoother, regardless of the computer's choice of setting. In town, the RX-7 feels free and willing to rev a shade quicker than the 944 does.

All this extends to the open road, as in ride and noise. Neither car has surplus power in 5th gear, by the way, as both are geared to get maximum economy within reason. On an interstate highway with a long upgrade, the 944 needed nearly full throttle to maintain 70 mph into a strong head wind, while the RX-7's driver found it kinder to leave the gearbox in 4th. Road warriors, these cars aren't.

Both have relatively wide ratios in their gearboxes; the 944 has 4th and 5th fairly high, with a gap between 3rd and 4th, while the RX-7's jump is between 4th and 5th. Both have mid-range torque and there's a gear for every occasion. But the sporting natures of both don't appear until the going gets curvy.

Through the twisty bits, both are excellent, rewarding the driver with control and speed. Both are sure of foot. The as-sold tires worked well in the dry and where there'd been rain shortly before our arrival. Sorry, no rain or snow was included in the trip itself.

The Mazda did give one hint of things to come. On a straight but crowned desert road, wavy with dips, the RX-7 was busy at 100 mph. Not tricky, not unstable, but demanding constant driver input and attention as the chassis seemed to move about more than the speed or road surface called for.

GENERAL DATA

	Mazda RX-7 GXL	Porsche 944
Base price	$16,645	$22,950
Price as tested[1]	$17,954	$29,682
Body type	coupe	coupe
Layout, engine/drive	front/rear	front/rear
Curb weight, lb	2790	2900
Test weight	2950	3060
Wgt dist, f/r, %	50/50	49/51
Wheelbase, in.	95.7	94.5
Track, f/r	57.1/56.7	58.2/57.1
Length	168.9	170.0
Width	66.5	68.3
Height	49.8	50.2
Engine type	2-rotor Wankel	sohc inline-4
Bore x stroke, mm	80.0 x 240.0[2]	100.0 x 78.9
Displacement, cc	1308	2479
Compression ratio	9.4:1	9.7:1
Bhp @ rpm, SAE net	146 @ 6500	147 @ 5800
Torque @ rpm, lb-ft	138 @ 3500	144 @ 3000
Fuel injection	Mazda electronic	Bosch LE-Jetronic
Transmission	5-sp manual	5-sp manual
Gear ratio, :1	3.48/2.00/1.37 1.00/0.71	3.60/2.13/1.46 1.07/0.73
Final drive ratio, :1	4.10	3.89
Steering type	rack & pinion, elect. variable pwr assist	rack & pinion, pwr assist
Brake system, f/r	10.9-in. vented discs/ 10.7-in. vented discs; vacuum assisted	11.1-in. vented discs/ 11.4-in. vented discs; vacuum assisted
Wheels	alloy, 15 x 6JJ	alloy, 15 x 7J
Tires	Bridgestone Potenza, RE71, 205/60VR-15	Pirelli P6, 215/60VR-15
Suspension, f/r	MacPherson struts, lower A-arms, coil springs, tube shocks, anti-roll bar/ trailing arms with camber-control arms on drop links, semi-trailing links, floating hubs, coil springs, tube shocks, anti-roll bar	MacPherson struts, lower A-arms, coil springs, tube shocks, anti-roll bar/ semi-trailing arms, torsion bars, tube shocks, anti-roll bar

[1]Price as tested includes: For the RX-7 GXL, leather seats ($720), 2+2 rear seats ($500), floor mats ($49), center armrest ($40); for the 944, touring package ($2715), 15-in. alloy wheels ($744), AM/FM stereo/cassette ($625), limited-slip diff ($595), sport seats ($590), sport shock absorbers and anti-roll bars ($295), central locking ($250), misc options ($340)

[2]Bore width x chamber major axis, mm

Taken to the limit

THE HARSHEST test in this comparison, and the most revealing, was on the track, namely Willow Springs International Raceway. This isn't always part of the comparison formula. For another class of car, perhaps, lap times would be useful and informative but not truly pertinent.

Willow Springs is a good equalizer. It's fast enough to keep the contest out of the slalom category, yet there isn't a long, fast straight, so power won't make up for shortcomings elsewhere.

For plain results in this present case, the 944 was faster around the track than the RX-7. The difference worked out to 2.6 seconds, best laps, same driver, same day.

But lap times didn't tell the whole story. Both cars were also clocked through the trickiest part of the track, from the top of the keyhole down a reverse camber right plunging into a banked left, and from the crest beginning the back straight to the braking mark-

Mazda's 2-rotor Wankel resides beneath induction plumbing.

The right half of Porsche's 928 V-8 powers the 944.

er heading into Turn 9, which Dan Gurney says is the most demanding corner of his experience.

The 944 and RX-7 were dead equal in the slow keyhole section, but the 944 made up time as the road (and cars) opened up; mere fractions of a second, but enough to make a difference.

Power to weight and steady-state cornering, remember, are close, and so are top speeds measured with a long, straight run. The 944 did better on the track because it could be driven faster with less drama on the demanding, faster corners.

The Porsche goes where it's pointed. Response to input is precisely linear. The harder you steer, the more the 944 turns. Mild understeer is there all the time, except on slow corners where power will get the tail out and around and back under control on command. The 944 tracks straight and does just what the driver wants. Even on the fast parts, the firm suspension that intrudes on highway travel keeps all four wheels exactly where they should be. The 944's reactions are clear and unmuffled right up to its limits.

The RX-7 doesn't work as well, not on the limit.

This can be described in two poetic ways. First, it's as if the Mazda engineers have built two stages into the suspension. The early stage, the one in which the settings are soft, is there up to perhaps 4/10s of maximum. Then the second stage takes over, the

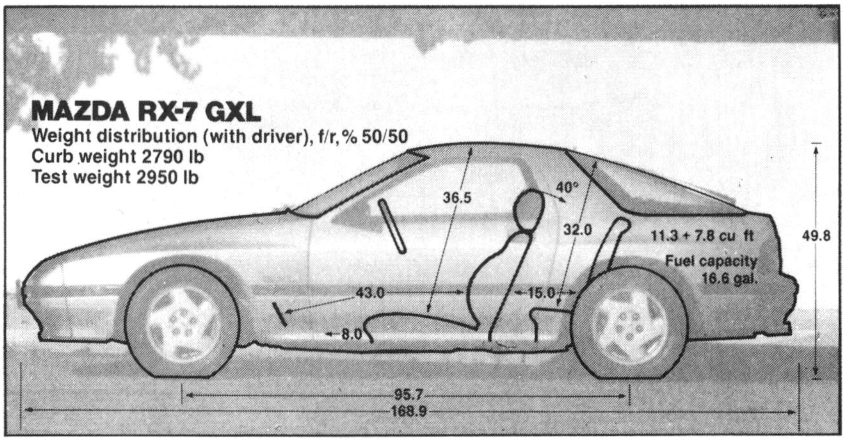

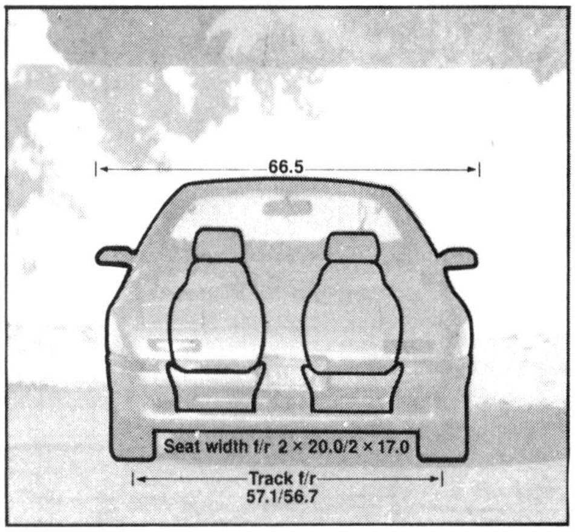

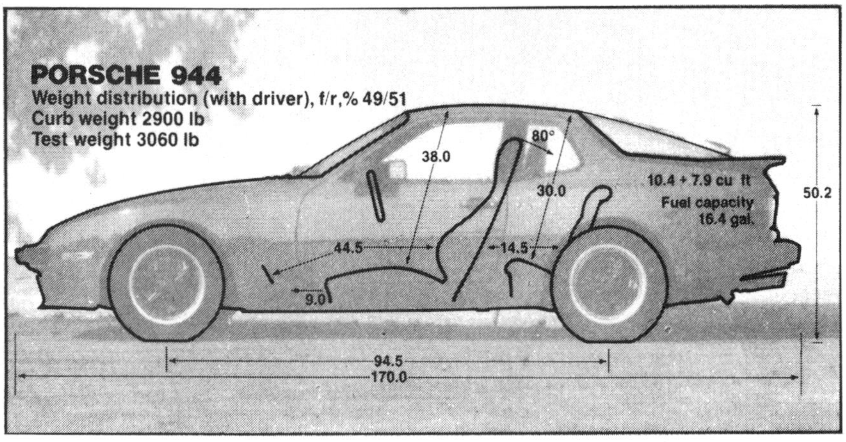

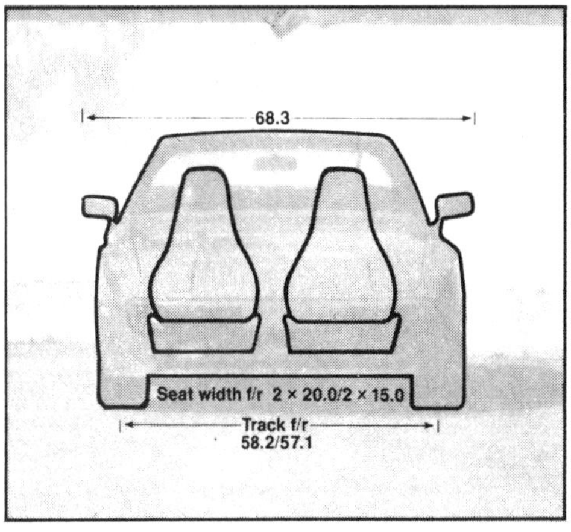

firm settings with the rear wheels steering into the turn while a few tenths of a "g" ago they were steering away from it. The two fail to feel completely in tune, or maybe all the adjustments made to one don't work with the other.

The second description is to think of the car as two rectangles, a soft flexible one bounded by a firm one. You, the driver, are inside the first limit and when you steer, brake, accelerate and so forth, you can feel the inner boundary moving within, up against and then away from the outer boundary.

What this means mostly is that when the RX-7 is hammer down, running flat-out on the inside edge of the track and the driver prepares to clip the apex *here* and let the car drift out until it's time to correct gently and line up for full braking to set up for the entrance to the last turn, flat-out again, what crosses the driver s mind is the sensation of how much automation, computerization and servo action is in between him and the steered wheels.

This brings us back to the RX-7's suspension, steering and rear-wheel geometry. The overall design and the optional extras were engineered to give good response and control at the extremes of expected conditions.

This in most instances to most drivers is described as good handling. Steering is light and sure; the chassis sets itself against the limit of initial travel and roll. In itself, no problem; if measuring the new RX-7 against the old version, a big improvement.

Nor is performance at the edge of sporting behavior the only factor in a comparison. Others are shown in our Subjective Evaluations, with the tally nearby.

The votes for some of these qualities deserve mention.

Exterior looks, for instance, came down to personal appeal of each car, plus the belief that the Porsche by golly looks like a Por- ➳

PERFORMANCE

	Mazda RX-7 GXL	Porsche 944
Acceleration:		
Time to distance, sec:		
0–100 ft	3.3	3.4
0–500 ft	8.9	9.1
0–1320 ft (¼ mi)	16.5	16.6
Speed at end of ¼ mi, mph	85.0	85.0
Time to speed, sec:		
0–30 mph	2.7	2.8
0–60 mph	8.5	8.9
0–80 mph	14.6	14.6
Top speed, mph	119	123
Willow Springs lap time, min: sec	1:48.8	1:46.2
Trip fuel economy, mpg	18.0	22.0
Brakes:		
Stopping distances, ft, from:		
60 mph	151	145
80 mph	267	256
Pedal effort for 0.5g stop, lb	22	19
Fade, % increase in effort,		
6 stop from 60 mph @ 0.5g	nil	nil
Handling		
Lateral acceleration, g	0.83	0.86
Slalom speed, mph	62.0	62.5
Interior noise, dBA		
Idle in neutral	53	59
Maximum, 1st gear	73	76
Constant 30 mph	67	65
50 mph	71	68
70 mph	74	72

CALCULATED DATA

	Mazda RX-7 GXL	Porsche 944
Lb/bhp (test weight)	20.2	20.8
Bhp/liter	111.6	59.3
Engine rpm @ 60 mph, (5th gear)	2450	2250
R&T steering index	0.87	1.25

PHOTO BY STEVE KIMBALL

PHOTO BY JOHN LAMM

Up to Spec?

IN THEIR DEVELOPMENT of the new RX-7, Mazda designers had one target firmly in mind, the Porsche 944. It was used in early development work and Mazda had one at the Seattle long lead for comparison drives, on the road and on the track. A comparison test between the two was a natural and in this comparison we picked the RX-7 model closest in specifications—the upmarket GXL—to its German rival. However, we were not all that happy with some of the Mazda's behavior at Willow Springs where it exhibited rear brake locking and some unsettling twitchiness in a couple of the faster turns.

What to do? We called Mazda and, without disclosing either the track times or our conclusions, we asked them to look at the car. The verdict? We had tested an early production car and some of the chassis pieces might not be to spec, according to Mazda engineers. To test this theory, we were given another GXL, coincidentally minus the rear jump seats, and we returned to Willow Springs. This car exhibited virtually identical behavior and the average of the two best laps was 0.5 seconds faster than the first, certainly within reason for two cars that were as identical as production-line assembly allows and tested by the same driver on the same track but at times two weeks apart and under different ambient conditions.

Totally unexpected was the Porsche's improvement. The *same* car was 2.7 sec quicker than previously. From the first lap the driver knew the Porsche was faster. He could sense it and he could read it on the tach at the exit of every corner.

How to account for this seeming anomoly? Break-in miles. When the 944 first went to Willow it had roughly 600 miles on its odometer. Two weeks later the odo read almost 2000 miles and the engine felt noticeably freer. "Finding" 2.7 sec is a lot of time at Willow Springs, but it points out in dramatic fashion how much difference a well broken-in and a well tuned engine can make. You could also bet money on the 944's acceleration times being quicker than those listed in this comparison test. We'll update those in the near future; scheduling conflicts prevented us from keeping the 944 long enough to retest the Porsche for this issue.

So much for the 944. Let's get back to the RX-7. Mazda was perplexed by our results and concerned enough to fly Takaaki Ito, manager of chassis design department number 3, and a couple of his assistants to the U.S. to check the two GXLs we'd tested. They spent a few hours with us discussing our findings and clarifying the design goals of the new RX-7.

After testing and tearing down both cars, Ito discovered that some of the suspension bushings in both cars weren't to spec and asked us to make one more retest. We agreed, requesting at the same time a model equipped with the optional Sport Package and larger disc brakes for comparison. We were interested in learning whether this lighter RX-7 without the GXL's computerized suspension might exhibit more predictable handling.

We were gratified by the performance of both cars. The GXL (same car as for our first retest) was more than a second faster around Willow and, more important, more predictable and secure feeling around the fast turns. So apparently its non-spec bushings had their detrimental effect. But it still suffered from erratic rear locking, which required the driver to exercise considerable caution when cornering with the brakes on. The Sport Package RX-7, equipped with the same brakes, didn't suffer from this problem. This leads us to believe that it was a problem with only the two GXL models because of test abuse prior to any of our testing and not characteristic of all RX-7s equipped with these brakes. With more consistent braking behavior and a lighter curb weight, the Sport Package car was 0.9 sec quicker than the GXL model. It also had more responsive steering and better turn-in with identical Bridgestone tires.

What have we learned from all this? Simply, for 99 percent of U.S. drivers for the 99 percent of their normal driving, the GXL offers comfort, performance and equipment closest to the 944. For those 1 percent who want to knock heads with the Porsche in showroom stock racing, the lighter-weight base RX-7 equipped with the optional Sport Package and larger discs is the hot ticket.—*John Dinkel*

COMPARATIVE LAP TIMES

	Mazda RX-7	Porsche 944	Porsche/Mazda Differences
GXL 1	1:50.2	1:48.9	1.3
GXL 2	1:49.7	1:46.2	3.5
GXL 2/Sport	1:48.8/1:47.9		2.6/1.7

COMPARATIVE TEST DATA

	RX-7 GXL	RX-7 Sport	944
0–60 mph, sec	8.5	8.1	8.9
Lateral acceleration, g	0.83	0.84	0.86
Slalom speed, mph	62.0	64.3	62.5

sche (admittedly a latter-day one), while the RX-7 has no family tree or perhaps it has too many ancestors from various places to establish a firm identity. Against that, the Mazda studio did a better job integrating the several items, front bumpers, for example, than the Porsche studio did.

Both cars get full marks for analog instruments. Dials with needles in them are easier to read at a glance than digitals. The RX-7's red on black, though, is harder to read in full daylight than the Porsche's higher-contrast white markings.

The Mazda has a larger door and the windshield is farther away from steering wheel and seat edge, while the seat itself is shallower, so the occupants have better egress.

Both cars have rear somethings that cannot be described as seats. Adults cannot sit there—period. The RX-7 gains here only because the front seatback can be released with one hand, and the seat slides forward for easier access. Both are hatchback coupes, both have remote release and equally useful luggage space.

The engines are nearly a draw, with the only real advantage going to the Porsche because it delivers the same power on less fuel, or to the Mazda because it's got the same power for less noise; take your pick.

Far as that goes, all categories were left unweighted, 1 through 10, terrible to wonderful, with no adjustment for how much importance any one person might put on any one factor. If egress or seating comfort is more important than exterior finish or looks, that's left to the individual who so believes.

The same is partially true in the preference column.

The real question here is value, perceived or realized.

Along with raised sights, Mazda's marketing plan includes a different structure. Where the earlier RX-7 was sold in a series of combinations, as in you couldn't get the more powerful engine without also getting the luxury package, the new model comes with the same engine no matter what else is on the car. The top of the RX-7 line is the car shown here, the GXL, which comes with larger wheels and tires and the sports suspension, plus power assisted steering and adjustable shocks plus again the luxury items like sunroof. So the car tested has a suggested retail price of about $18,000 against the Porsche's sticker of almost $30,000.

Now both of these cars are rolling showcases of their manufacturers' goods. And, after the fact, it's fair to wonder if perhaps the plain RX-7 with sports package, the wheels, tires and firm suspension, but without the power assist or semi-automatic suspension wouldn't be better under some conditions and some $4000 cheaper in the (couldn't resist) bargain. Also, a quite capable 944 can be had without investing $6000 in options.

In any case, value is determined by who writes the check. Because one can get a better cost/performance ratio from, say, an MR2, a CRX, a V-8 Mustang or Camaro, the buyer of either the 944 or RX-7 is paying for something besides speed.

Sorry to sound elitist, but a Porsche is a Porsche. It brings status, never mind how the 911 owner may refuse to return waves. The 944 likely will hold its value. None of this will matter to the potential buyer who can swing Mazda's entry fee, but not the Porsche's, nor to the buyer who must have a Porsche because it's a Porsche.

But for those in between, in the long run, all intangibles considered, the 944 looks to be a better value.

Finally, for the car versus car, repeating for emphasis, the shortcomings of the RX-7's scientific, computer-assisted steering and suspension are only apparent on the ragged edge, while they do improve the car the rest of the time. Mazda's engineers did a textbook job of designing the car to do what most people expect of it.

Porsche's engineers must have driven their projects not only on the test track, but also on the race track and on the road, and they made the car behave best at its limits, even if that meant taking a toll on the driver the rest of the time.

The Mazda RX-7 is easier to drive easily, while the Porsche 944 is easier to drive hard.

So the 944 is still the better sports/GT car, and Porsche continues to pay what Cadillac used to call the penalty of leadership, i.e., the 944 is still the target.

CUMULATIVE RATINGS–SUBJECTIVE EVALUATIONS

	Mazda RX-7 GXL	Porsche 944	Comments
Performance			
Engine	6.3	**7.3**	Neither has major weakness; Porsche develops better torque.
Gearbox	**7.3**	7.0	Mazda has lighter touch; Porsche relates an inertial feel.
Steering	6.5	**9.0**	944 feel is terrific; RX-7, remote and overboosted.
Brakes	6.5	**8.0**	Porsche's are powerful; Mazda's are spongy and less predictable.
Ride	**7.0**	6.8	944 has improved, but RX-7 is very good for a sports car.
Handling	7.0	**9.0**	Stuttgart's got the balance right; Hiroshima's still looking.
Body structure	**7.0**	**7.0**	A draw—both are virtually unflappable on rough roads.
Average	6.8	**7.7**	
Comfort/Controls:			
Driving position	7.8	**8.3**	Both good, the Porsche steering wheel and pedals fit better.
Controls	6.5	**7.8**	Long clutch travel, some dash controls hidden by steering wheel.
Instrumentation	6.3	**8.3**	Mazda's gauge markings are hard to read.
Outward vision	6.8	**7.0**	RX-7's big B-pillars can be a hindrance.
Quiétness	**7.3**	6.5	Porsche engine vibration makes for a loud idle.
Heat/vent/air cond	**7.8**	7.3	The Mazda controls are slightly more logical.
Ingress/egress	**7.3**	6.8	RX-7 doors open wider; 944 seats are more obstructive.
Front seat	7.3	**8.5**	Porsche's have just the right shape—provided you're of average size.
Rear seat	**1.8**	0.8	Both are for token use only; Mazda's have easier access.
Luggage & loading	**6.8**	6.0	Mazda cargo floor is deeper.
Average	6.6	**6.7**	
Design/Styling:			
Exterior styling	6.8	**7.8**	Though evolutionary, the Porsche looks right; Mazda is derivative.
Exterior finish	**8.5**	7.0	944 shows its 924 heritage here & many wide body panel gaps.
Interior syling	6.5	**7.8**	No fault of the Japanese, the German look is more impressive.
Interior finish	**7.5**	**7.5**	High quality in both, though the 944's higher price shows.
Average	7.3	**7.5**	
Overall average	6.8	**7.2**	
Staff preference[1]:			
Price-independent	4	**8**	
Price dependent	**7**	5	

[1]four staff members: 1st choice, 2 points, 2nd choice, 1 point.

MAZDA RX-7 TURBO

Hard on the heels of the new RX-7, the Turbo version ups the power ante

PHOTOS BY JOHN LAMM

mazda WHAT POWER REALLY does, never mind what they told you in school, is attract. For as long as people have been driving for fun—probably as long as people have been driving—it's been more fun if the car has more power. Or at least so think a lot of folks.

So when Mazda's designers began planning the new RX-7, one of the plan's segments was a turbocharged version of the car. Turbocharging is an easy way to get more power from the same basic engine; the public knows Turbo means performance and, thanks in part to the hard lessons pioneering firms learned and passed on to the rest of the world, the drawbacks are well-known and (lately, anyway) pretty well taken care of.

Mostly, though, the Turbo version was seen by Mazda as a model that could put the RX-7 right up there, ready to compete against sports cars that cost more, had larger engines or both.

The RX-7 Turbo project, called Turbo II because this is the second turbocharged rotary engine Mazda has built, began with the 13B twin-rotary powerplant used in the other new RX-7 models. Accompanying changes were minor: The compression ratio was reduced from 9.4 to 8.5:1 so that the added pressure from the turbo wouldn't mean more pressure than the engine can handle, and the spark curve and fuel calibration were revised to suit the turbo's characteristics.

The turbo unit itself is positioned close to the engine's exhaust ports. There's a subtle benefit here. In some ways a Wankel rotary is like a 2-stroke, mostly because it doesn't have valves. So the exhaust gas has a clean exit through the ports and can blast directly against the turbo's driven turbine wheel. And the closer the driven wheel is to the exhaust ports, the less energy is lost. The

AT A GLANCE	Mazda RX-7 Turbo	Chevrolet Corvette	Porsche 944 Turbo
Curb weight, lb	est 2830	3280	3000
Engine/drive	2-rotor Wankel/rwd	V-8/rwd	inline-4/rwd
Transmission	5-sp M	4 + OD M	5-sp M
0–60 mph, sec	6.6	5.8	6.0
Standing ¼ mi, sec @ mph	15.2 @ 92.5	14.4 @ 96.0	14.6 @ 97.5
Stopping distance from 60 mph, ft	154	133	143
Lateral acceleration, g	0.85	0.91	0.90
Slalom speed, mph	63.1	61.0	62.8
Fuel economy, mpg	19.2[1]	19.0	19.4[1]

	Pro	Con
RX-7 Turbo:	smooth, high-revving rotary engine, high performance, comfortable and practical interior	styling lacks originality, low fuel economy for performance and weight, lacks ABS
Corvette:	very high performance with reasonable fuel economy, very high cornering power, ABS	bulky for a 2-passenger car, tends to rattle and squeak, awkward operation of 4 + OD transmission
944 Turbo:	very high performance with reasonable fuel economy, excellent handling, top workmanship	meager standard equipment for price, turbo lag, lacks ABS

[1]trip fuel economy

turbo is shielded from the engine's own, unwelcome heat, and the bearings are water-cooled; one of the lessons learned by the pioneers was that the turbocharger bearings live a hard life indeed.

Called a "twin-scroll" turbocharger, Mazda's has dual turbine inlet chambers. The primary chamber is small so that exhaust gas from the engine at low engine speeds can strike the turbine blades at an efficient angle, actually almost right angles. The secondary chamber is larger, with a more obtuse angle of attack on a different portion of the turbine. This split chamber thus acts as a variable nozzle. Meanwhile, our trusty friend the microprocessor is reading engine speed and air volume in the intake manifold and opening or closing a valve on the secondary chamber. The hoped-for effect is that of a small turbocharger at low engine speed and a larger one at higher speeds, and a reduction in turbocharging's classic lag.

Rated power and torque of the Turbo engine are 182 bhp and 183 lb-ft, versus 146 and 138 for the standard 13B engine: substantial increases, due in part to the innovative turbocharger and in part to an intercooler. Incoming air is routed through the intercooler, which is mounted above the engine, and that means the Turbo gets an impressive (and productive, by all measures) hood scoop. Not only does it work, it's an easy way to spot the Turbo RX-7 from afar; unlike some other factories we could mention, Mazda hasn't hung a flurry of labels and badges all over the car.

Aft of the engine goes a larger, stronger clutch and a larger, stronger transmission, both for easily understood reasons. The gearbox's ratios change fractionally, as the various gear wheels are larger so the math doesn't come out quite the same, but in effect the spreads between 1st and 2nd, 2nd and 3rd, etc, are about the same. The final drive ratio is 3.91:1, shared with the automatic-transmission RX-7s and a bit taller than the 4.10:1 of the unblown 5-speed cars. There is no automatic option with the turbo.

Because Turbo buyers are presumed to be sporting types who'll use the extra power, as in drive faster, the Turbo comes with its suspension preset on Full Hard, as opposed to the driver-adjustable suspension on the unblown GXL. The electronically controlled power-assisted steering is the same (but, oddly, optional), and naturally the Turbo has the programmed toe-in, toe-out rear suspension of the other models.

The Turbo comes with 16 x 7-in. alloy wheels and 55-section tires (the standard RX-7 has 14 x 5-in. wheels and 70-section tires, the GXL 15 x 6-in. and 60-section); lower and less flexible and, therefore, better handling is the reason here. And the Turbo's rear disc brakes are vented, again in the expection of more speed bringing more braking demand. Mazda generally builds a series of complete equipment steps, as in you can't get the optional suspension without the optioned interior and vice versa. Our test Turbo had manual window cranks, mirrors and seats, along with full-boat sound system and air conditioning; but the production model will come equipped pretty much like the upmarket non-turbo GXL, with electric windows, electric mirrors and most of the other high-zoot items.

The important part, as in Turbo vs GXL or even the rest of the sports-car competition, is easily expressed in numbers: The Turbo is quicker to any speed and distance than the plain GXL. Runs from 0–60 mph take 6.6 sec against 8.5, for example, and the standing quarter mile arrives in 15.2 sec rather than 16.1. It's all relative, of course, and there are quicker cars on the market. But the Turbo romps away from ordinary sedans and its performance puts it into the truly fast category, which is what the designers had in mind. On a mostly open-road test trip the RX-7 Turbo turned in 19.2 mpg—respectable, but not impressive in view of the car's performance and weight.

There are some shortcomings. Perhaps reading Mazda's book with the impressive details of engineering raises one's expectations. Even so, the Turbo (like most turbo engines) has a lower compression ratio than the plain 13B and it's geared higher, which adds up to less surplus power when the car is cruising at constant, nearly closed throttle, at any road speed.

When the driver mashes down on the long thin pedal, the Turbo's boost gauge jumps quickly to positive, just as the book says it should.

But the car doesn't move as quickly as the needle. At low rpm and low boost, there isn't extra power and it takes some time for the engine to respond. This isn't classic turbo lag; it's more like a concave power curve. The Turbo II rotary, like your typical 2.0-liter turbocharged 4-cylinder engine powering nearly 3000 pounds of car, needs to be high on the tach and deeply into the throttle before things get interesting. It's a normal turbocharger installation, in other words, while on paper it promised to be more than that.

Next, the extra pressure and torque show up as added noise and harshness, a howl from the gears and vibration transmitted through the drivetrain to the cockpit, all of which isn't there in the plain RX-7. Again, what one usually sees in the first year of a turbo option, which is just what this is.

This extends to the suspension. Full Hard, remember. The ride is firm on the smoothest surfaces and harsh to the point of jiggly on dots, cracks and bumps. The low-profile tires were picked on the basis of stick, which they do. What they don't do is absorb road irregularities.

The microprocessed steering assist is programmed to be light and quick under all conditions. This, it is. The Turbo turns quickly and reacts to driver input as fast as the driver can put it in. But it happens with a certain detachment, a masking of road feel in the interest of effortless driving.

Something of the same could be said for the rear suspension, the one that toes the ouside rear wheel out under a pre-determined loading, then in if the loading goes to the next stage. And the rear wheels steer, sort of, under full braking or full acceleration.

This programming works fine, provided the driver is doing what the designers intended, no more and no less. The RX-7 Turbo is a fast car and handles well up to (presumably) the limit of what the engineers had in mind.

Beyond that, or if conditions are less than ideal, there's a lack of communication. The chassis is going to do what it's supposed to do, and the driver must be prepared and go along with the program; with a more conventional system it might be the other way.

Further, although the new rear suspension is heaps better than that of the old RX-7, the extra power and tires and brakes can overcome it. The rear gets light under full braking, and the wheels will judder and tend to lock on bumps going into a corner.

As a result, the RX-7 Turbo is easy to drive fast, much more difficult to drive faster than that, and demanding to the point of disquiet right on the edge.

Sounds terrible? Not that bad. The new RX-7 Turbo with performance suspension reminds one of the new-in-1984 Corvette with performance suspension. The engineers worked wonders in getting handling on billiard-table test tracks and learned from the customers that the real world isn't like that. What the RX-7 needs here is more development.

The rest of the RX-7 Turbo is RX-7, which is mostly useful, pleasant, comfortable, etc. The manually adjusted seats give enough cornering support and easy egress; the Turbo's slightly smaller steering wheel helps, and the brace for the left foot is welcome. There's a lot of luggage space, accessible through a generous hatch. All needed instruments are there, hampered only by an orange color scheme that is ineffective under some lighting conditions. The air conditioning and heating work fine, and so does the radio except maybe for all those tiny buttons—but that's the way they all are nowadays.

Enough digression.

The subject is Mazda's turbo, as seen on the RX-7, and the clear verdict is that it works. The car is faster than the plain version, fast enough to contend with just about anything on the road. The price in money seems reasonable; the price in noise and vibration, well, how attractive is power to you?

PRICE

List price, all POE	est $19,500
Price as tested	est $19,980

Price as tested includes std equip (air cond, AM/FM stereo/cassette, rear wiper), pwr asst strg (est $300), alarm system ($180); some planned std items (elect. windows and mirrors) not on test car

GENERAL

Curb weight, lb/kg	est 2830	est 1285
Test weight	est 2990	est 1355
Weight dist (with driver), f/r, %		50/50
Wheelbase, in./mm	95.7	2430
Track, front/rear	57.1/56.7	1450/1440
Length	168.9	4290
Width	66.5	1690
Height	49.8	1265
Trunk space, cu ft/liters	11.3+7.8	320+221
Fuel capacity, U.S. gal./liters	16.6	63

ENGINE

Type		turbocharged 2-rotor Wankel
Bore width x chamber major axis, in./mm	3.15 x 9.45	80.0 x 240.0
Displacement, cu in./cc	79.8	1308
Compression ratio		8.5:1
Bhp @ rpm, SAE net/kW		182/136 @ 6500
Torque @ rpm, lb-ft/Nm		183/248 @ 3500
Fuel injection		Mazda electronic
Fuel requirement		unleaded, 87 pump octane

DRIVETRAIN

Transmission	5-sp manual
Gear ratios: 5th (0.76)	2.97:1
4th (1.00)	3.91:1
3rd (1.39)	5.43:1
2nd (2.02)	7.90:1
1st (3.48)	13.61:1
Final drive ratio	3.91:1

CHASSIS & BODY

Layout	front engine/rear drive
Body/frame	unit steel
Brake system	10.9-in. vented discs front, 10.7-in. vented discs rear, vacuum assisted
Wheels	alloy, 15 x 7JJ
Tires	Goodyear Eagle GT, 205/55VR-16
Steering type	rack & pinion, electronic variable power assist
Turns lock-to-lock	2.7
Suspension, front/rear	MacPherson struts, A-arms, coil springs, tube shocks, anti-roll bar/trailing arms with camber-control arms on drop links, semi-trailing links, floating hubs, coil springs

CALCULATED DATA

Lb/bhp (test weight)	16.4
Engine revs/mi (5th gear, 60 mph)	2620
R&T steering index	0.87
Brake swept area, sq in./ton	280

ROAD TEST RESULTS

ACCELERATION

Time to distance, sec:	
0-100 ft	3.2
0-500 ft	8.3
0-1320 ft (¼ mi)	15.2
Speed at end of ¼ mi, mph	92.5
Time to speed, sec:	
0-30 mph	2.4
0-50 mph	5.1
0-60 mph	6.6
0-70 mph	8.8
0-80 mph	11.0
0-90 mph	18.0

SPEEDS IN GEARS

5th gear (6500 rpm)	est 150
4th (7000)	119
3rd (7000)	88
2nd (7000)	61
1st (7000)	36

FUEL ECONOMY

Trip fuel economy, mpg	19.2

BRAKES

Minimum stopping distances, ft:	
From 60 mph	154
From 80 mph	259
Control in panic stop	good
Pedal effort for 0.5g stop, lb	18
Fade: percent increase in pedal effort to maintain 0.5g deceleration in 6 stops from 60 mph	nil
Overall brake rating	good

HANDLING

Lateral accel, 100-ft radius, g	0.85
Speed thru 700-ft slalom, mph	63.1

INTERIOR NOISE

Constant 30 mph	71
50 mph	70
70 mph	74

ACCELERATION

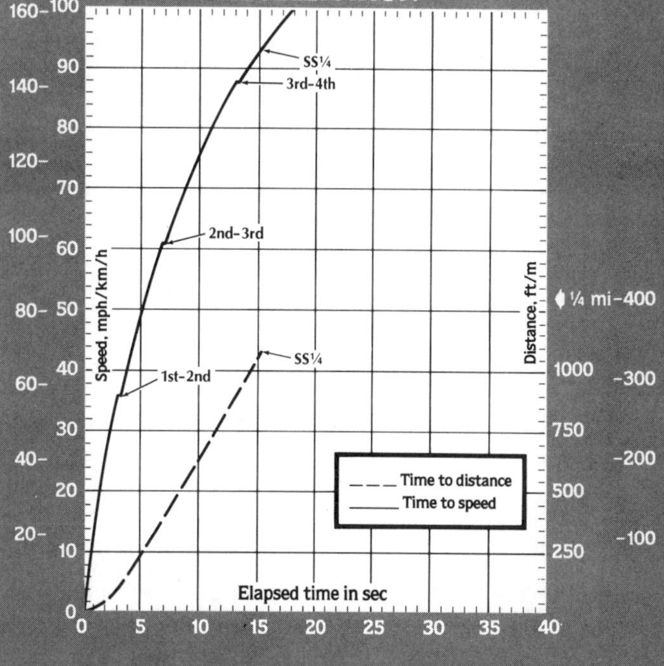

COMPARISON R&T ROAD TEST

CORVETTE VS 944 TURBO VS RX-7 TURBO

They're better than average, they're fast and they follow the same format. But that doesn't mean they're alike on the road.

PHOTOS BY JOHN LAMM

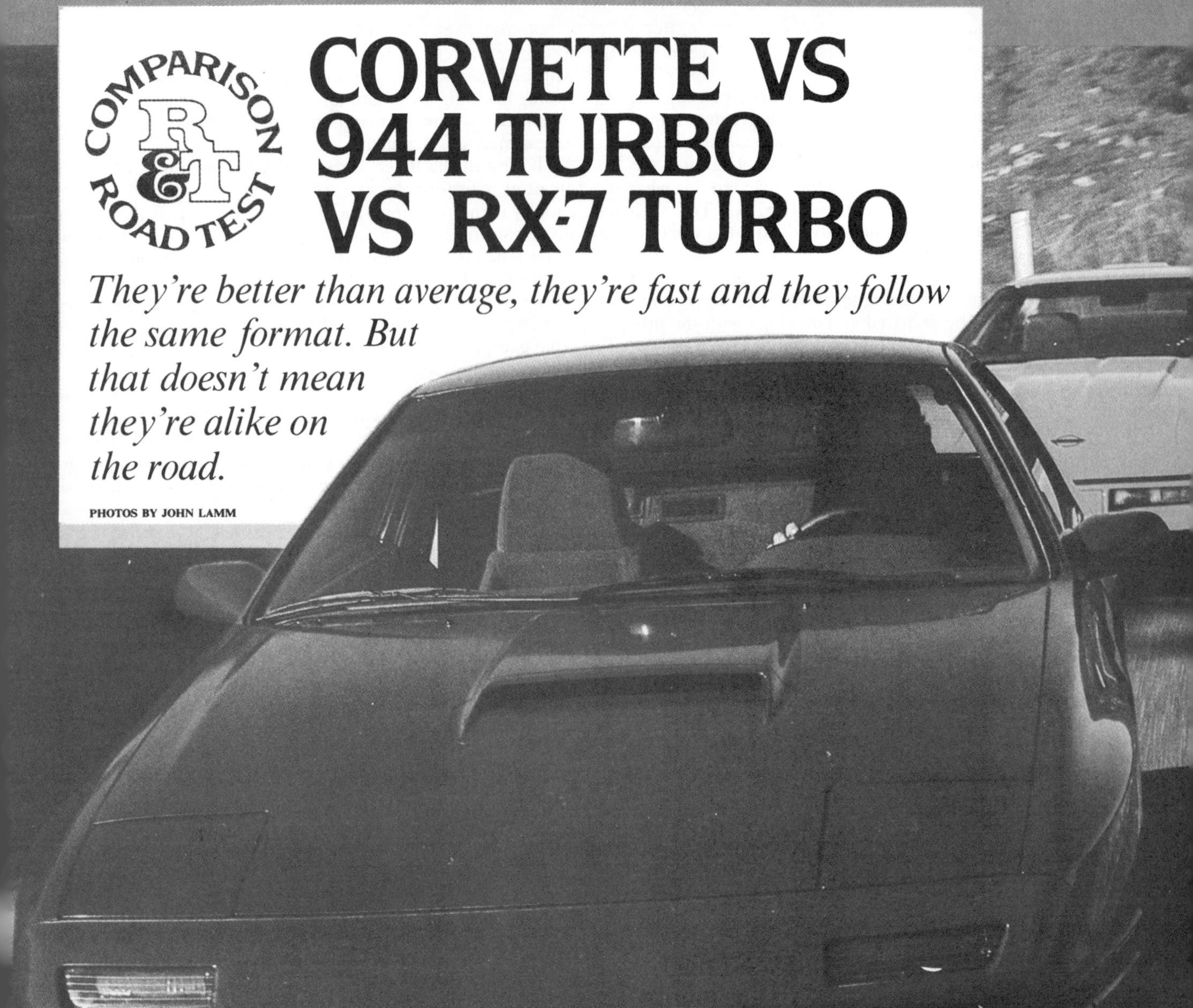

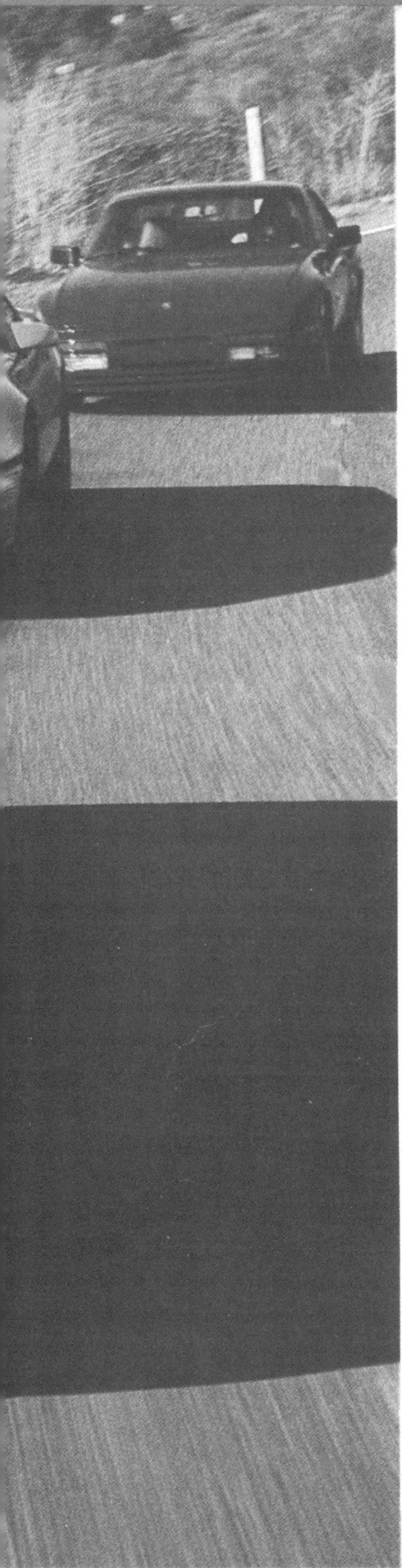

It is a truth universally acknowledged that a man in possession of a more than adequate income must be in want of a more than adequate car.

There is in consequence no shortage of such cars on the market. If adequate can be defined as reliable and practical transportation, then the surplus (in some cases, excess) to requirements can be obtained in a variety of ways, as in larger or faster or more elaborate, etc, than a car needs to be. All part of the entertainment.

If the extras are the entertainment, the field narrows. As in the present case, more than adequate becomes a car that travels from here to there with more speed, better performance, less fuss and less fatigue than an ordinary car can provide.

The three cars here will do that. They are also alike in that they are (effectively) 2-place sports cars with engines in front driving the rear wheels. Each in its own way represents a statement by the maker as to how to meet the demands of the demanding enthusiast.

And each in its own way is as markedly different from the other two as three cars of the same configuration can be.

In order of their appearance . . .

Chevrolet Corvette. The current model was introduced in 1984, as another Corvette in the model's format, i.e., big 2-seater with big V-8, highly tuned, with fiberglass body on steel frame. The current Vette is slicker and lighter than the previous version, and the 1986 is refined, relatively, compared with the 1984 and 1985 models. The 1986 car shown here benefits from the addition of anti-lock braking and from aluminum cylinder heads, now back in production after a short interruption (see R&T, February 1986). Price of the test car was on the high side because it had the optional suspension, leather seats, etc.

Porsche 944 Turbo. The 944 is a revitalized model, an upgraded and much better version of the ill-fated 924. The Turbo is the latest addition. It adds not just power, but power in the mid-range and on the top end, where it pays off in sporting surge and a really amazing top speed, albeit the speed isn't something one can use every day. Seen in a family group (R&T, January 1986), the 944 Turbo is the best Porsche in terms of fun per dollar and speed without stress.

Mazda RX-7 Turbo. The new RX-7 is new, a completely different car with the old name. In non-turbo form it's an impressive package and surely is the most elaborate sports car on the market: a rear suspension that toes the wheels in and out under cornering forces, in braking and during hard acceleration, and optional computer-controlled suspension that the driver can vary and that will adjust itself to what the chassis is doing.

The Turbo model, formally Turbo II (it's the second turbocharged Wankel engine Mazda has built; the first was sold only in Japan) is brand-new this instant, just off the boat. It's the most powerful version, and the most sporting, of what was already Mazda's leading light.

The turbocharger itself is as new and complex as is the car it attaches to. The engine is the 13B rotary seen in lesser RX-7s, with compression ratio reduced from 9.4 to 8.5:1 so it can accept the extra pressure from the turbo. The turbine is tucked up close to the engine's exhaust, for minimum lag, and it's shielded from heat and uses water-cooled bearings. There's an intercooler and the chamber around the driven side of the turbine has two compartments. At low rpm, the exhaust expands into the smaller chamber and spins the turbine quickly despite minimum flow. As the engine gains speed, the second half of the chamber is opened, so the turbine keeps pressure and accommodates the added incoming air. Further, the incoming blast is directed to different locations on the turbine blades at high and ➳

low rpm, for more efficiency. And, of course, it's all controlled by microprocessors, along with the spark advance curve, fuel mixture, etc.

The Turbo II engine brings with it larger wheels (16 x 7 in.) than those on the plain RX-7s. Like the Sport Package (and unlike the base and GXL models), the electrically adjustable shocks are not standard or available. The Turbo's settings are fixed at rates very similar to the adjustable system's firm level: Turbo drivers are assumed to be chargers. The test car gained a bit in the sticker department because the Turbo model doesn't come with the full luxury package; Turbo drivers are assumed to prefer performance to luxury. The car here forced its occupants to actually crank the windows up and down by hand, the brute.

These three cars are close in power to weight ratios and all three are designed for the same sort of use—they are all sports cars. And they are intended for daily use, assuming most buyers aren't so far beyond adequate income that they can have a stable filled with cars.

So the 2-seat trio was used for daily driving and for fun on sporting roads, and then scored on the basis of both.

The figures

As the clocked and timed data show, these are fast, quick and stable cars.

The Corvette is the quickest, because it has the most power and the most torque and both are available on demand, right off the line. Rev the engine, drop the clutch and bang! there you are at the other end of the drag strip. The power curve, if it can be called that, is so nicely matched to the torque and gearing that the Corvette

GENERAL DATA

	Chevrolet Corvette	Mazda RX-7 Turbo	Porsche 944 Turbo
Base price	$24,403	est $21,000	$29,500
Price as tested[1]	$27,493	est $22,000	$30,820
Body type/seating	coupe/2	coupe/2	coupe/2+2
Layout, engine/drive	front/rear	front-mid/rear	front/rear
Curb weight, lb	3280	est 2830	3000
Test weight	3450	est 2990	3145
Wgt dist, f/r, %	51/49	est 50/50	50/50
Wheelbase, in.	96.2	95.7	94.5
Track, f/r	59.6/60.4	57.1/56.7	58.2/57.1
Length	176.5	168.9	168.9
Width	71.0	66.5	68.3
Height	46.7	49.8	50.2
Seat width, f/r	2 x 20.0	2 x 20.0	2 x 19.5/2 x 15.0
Head room, f/r	35.5	36.5	37.0/32.5
Engine type	ohv V-8	turbo 2-rotor Wankel	turbo sohc inline-4
Bore x stroke, mm	101.6 x 88.4	80.0 x 240.0[2]	100.0 x 78.9
Displacement, cc	5733	1308	2479
Compression ratio	9.5:1	8.5:1	8.0:1
Bhp @ rpm, SAE net	230 @ 4000	182 @ 6500	217 @ 5800
Torque @ rpm, lb-ft	330 @ 3200	183 @ 3500	243 @ 3500
Fuel injection	Bosch multi-point	Mazda electronic	Bosch LE-Jetronic
Transmission	4-sp manual + OD[3]	5-sp manual	5-sp manual
Gear ratios, :1	2.88/1.91/1.33 1.00	3.48/2.02/1.39 1.00/0.76	3.50/2.06/1.40 1.03/0.83
Final drive ratio, :1	3.07	3.91	3.38
Steering type	rack & pinion, pwr assist	rack & pinion, elect. variable pwr assist	rack & pinion pwr assist
Steering ratio	13.0:1	15.2:1	22.4:1
Brake system, f/r	11.5-in. vented discs/ 11.5-in. vented discs; vacuum assist, ABS	10.9-in. vented discs/ 10.7-in. vented discs; vacuum assist	11.7-in. vented discs, 11.8-in. vented discs vacuum assist
Wheels	cast alloy, 16 x 9½	cast alloy, 16 x 7JJ	cast alloy, 16 x 7J front 16 x 8J rear
Tires	Goodyear Eagle VR-50, P255/50VR-16	Goodyear Eagle GT, 205/55VR-16	Pirelli P7, 205/55VR-16 front/ 225/50VR-16 rear
Suspension, f/r	unequal-length A-arms, transverse fiberglass leaf spring, tube shocks, anti-roll bar/upper & lower trailing arms, lateral arms, tie rods, halfshafts, transverse fiberglass leaf spring, tube shocks, anti-roll bar	MacPherson struts, A-arms, coil springs, tube shocks, anti-roll bar/trailing arms with camber-control links, semi-trailing links, floating hubs, coil springs, tube shocks, anti-roll bar	MacPherson struts, lower A-arms, coil springs, tube shocks, anti-roll bar/semi-trailing arms, torsion bars, tube shocks, anti-roll bar

[1]Price as tested includes: For the Corvette, std equip. (air cond, elect. window lifts, elect. adj mirrors), AM/FM stereo/cassette ($895), sunroof ($595), Z51 suspension ($470), leather seats ($400), elect. adj driver's seat ($215), cruise control ($185), central locking ($170), rear-window & mirror heat ($160); for the 944 Turbo, std equip. (air cond, elect. window lifts), sunroof ($695), AM/FM stereo/cassette ($625)

[2]Bore width x chamber major axis, mm.

[3]Automatic or elect. overdrive on gears 2–4; overdrive ratio is 0.67:1 compared with direct drive.

will also go very fast, and still spend very little of its life in the passing lane.

The 944 Turbo has the most top speed, mostly because it's slick and has a small frontal area and the engine does its best work at the upper end of the scale. It takes some time to hit the incredible top speed, relatively, because the turbo engine needs revs to get boost and until it's on boost it doesn't have surplus power. Turbochargers aren't made for the drag strip, but the 944 Turbo isn't slow; it will wax most any car it meets at the stoplight—except the plastic Chevrolet.

The RX-7 was something of a disappointment. Again, not a slow car by any standard—the Turbo II cuts useful seconds off the times of the non-turbo RX-7, itself no slug. The turbo does add power.

But it's odd. Mash the pedal and the boost gauge shows instant plus, just as the brochure says it will. But the acceleration isn't as quick to arrive. The turbo brings a taller final drive for quieter cruising, while the boost makes downshifting for hills unnecessary, but the result is an engine that doesn't make added power for the second (or seconds) it needs to spool up. It isn't turbo lag so much as it's power delay.

And when the car is launched off the starting line and the gearshift and clutch are worked the way the drag strip demands, the car leaps and judders. The otherwise cushy drivetrain is compressed and released under more power than it was expected to handle, or so it feels. The RX-7 Turbo has the sort of characteristics we more often see on a car modified by an outside firm, i.e., more power and speed and better suspension for fast driving, at the expense of comfort and with flaws that wouldn't have been there if the builders ➳

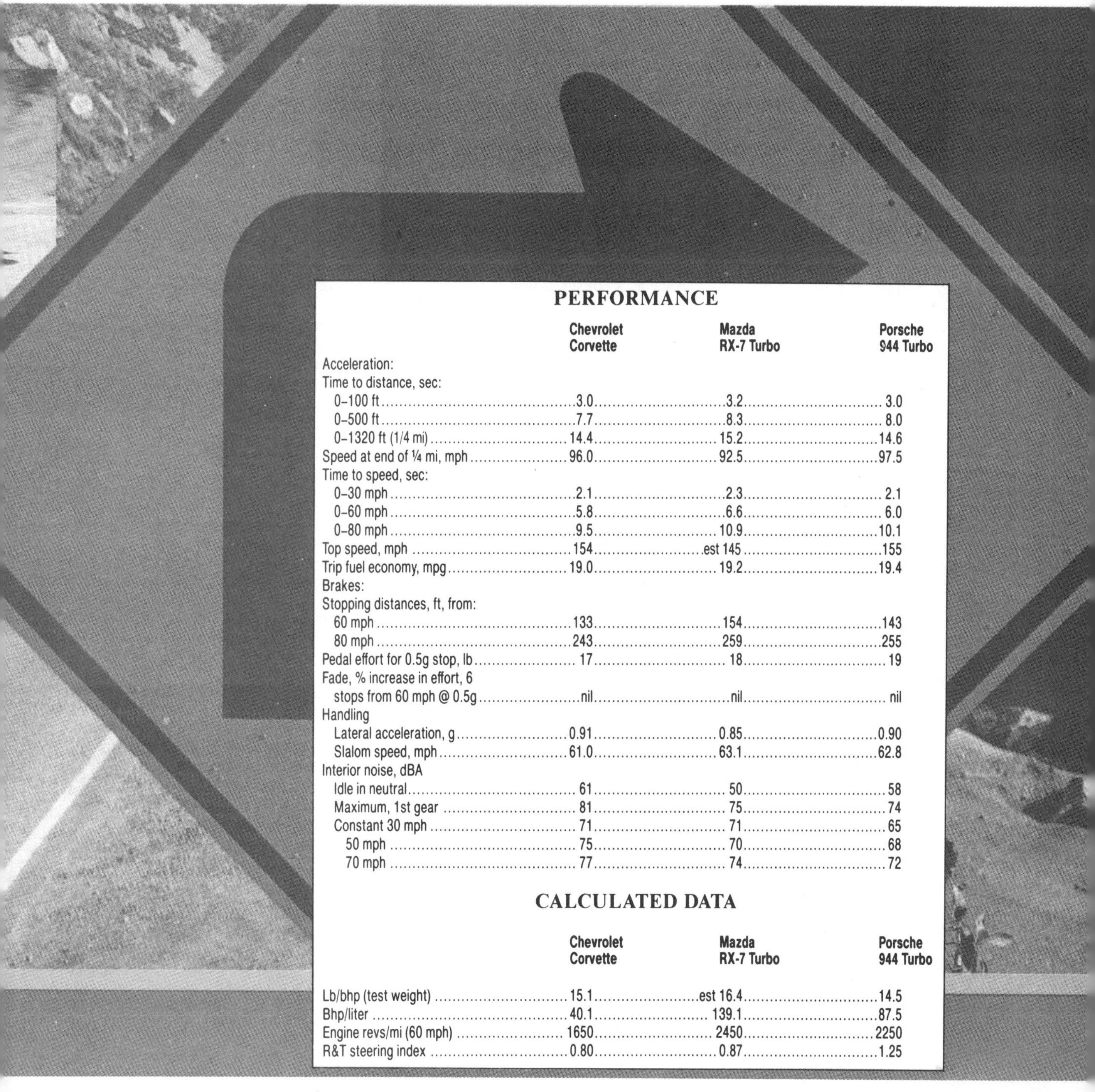

PERFORMANCE

	Chevrolet Corvette	Mazda RX-7 Turbo	Porsche 944 Turbo
Acceleration:			
Time to distance, sec:			
0–100 ft	3.0	3.2	3.0
0–500 ft	7.7	8.3	8.0
0–1320 ft (1/4 mi)	14.4	15.2	14.6
Speed at end of ¼ mi, mph	96.0	92.5	97.5
Time to speed, sec:			
0–30 mph	2.1	2.3	2.1
0–60 mph	5.8	6.6	6.0
0–80 mph	9.5	10.9	10.1
Top speed, mph	154	est 145	155
Trip fuel economy, mpg	19.0	19.2	19.4
Brakes:			
Stopping distances, ft, from:			
60 mph	133	154	143
80 mph	243	259	255
Pedal effort for 0.5g stop, lb	17	18	19
Fade, % increase in effort, 6 stops from 60 mph @ 0.5g	nil	nil	nil
Handling			
Lateral acceleration, g	0.91	0.85	0.90
Slalom speed, mph	61.0	63.1	62.8
Interior noise, dBA			
Idle in neutral	61	50	58
Maximum, 1st gear	81	75	74
Constant 30 mph	71	71	65
50 mph	75	70	68
70 mph	77	74	72

CALCULATED DATA

	Chevrolet Corvette	Mazda RX-7 Turbo	Porsche 944 Turbo
Lb/bhp (test weight)	15.1	est 16.4	14.5
Bhp/liter	40.1	139.1	87.5
Engine revs/mi (60 mph)	1650	2450	2250
R&T steering index	0.80	0.87	1.25

had known from the start how much power they'd be dealing with. From outside, it's understandable. From the factory, it's a puzzle.

Grading on curves

MOUNTAIN DRIVING in sports cars is a secret sort of art; here, the Corvette thrives. Not just because of the power but because the power is so *there.* Quick, massive, arm-stretching response, while the driver is cradled by four huge, sticky tires. The Corvette thumps and it bumps but it has an amazing ability to go fast, predictably, while always feeling well within the limits. Click into 4th overdrive and let it cruise. If power is needed, push down and the gearbox drops from overdrive into direct and there you are, enough power to hang the tail out, and to bring it back.

On the dreaded interstate the mighty V-8 is loafing while the instrument panel tells you at least as much as you want to know and the suspension works audibly; that is, there is rumble and shake over the bumps and dots.

The 944 Turbo is just as good, but in a different, more restrained, way.

First, the power is there but the driver has to find it, or line it up in advance, with higher revs.

Next, the steering is beautifully balanced, but not as quick. The Porsche tells the driver exactly what's happening and it rewards precise, controlled input. This communication, or interaction, works so well it's hard to imagine getting into trouble with the 944 Turbo, but the car does have smaller tires than its American rival, so while it covers ground with less fuss, it doesn't cover the ground quite so quickly.

Automation hasn't done the RX-7 all the favors its designers think. It's plenty stiff and has more than enough tire but because the power assisted steering is programmed to feel constant, and because the suspension is tuned to react to what's supposed to happen, man and machine will work at cross purposes unless both are doing exactly what the guys back in the lab thought they'd do. The RX-7 Turbo has light controls and everything works, so it's easy to drive fast. But pushed beyond fast, it works against itself and none of the panel drivers enjoyed the feeling.

Off boost, just going down the road wondering why all those people are clogging the fast lane, the impression is one of more gear and engine noise and more road shock fed into the cockpit than the plain RX-7 has.

Sum versus parts

ALL THREE cars get a passing grade. They all have more than adequate performance and can serve as the owner's only car or daily car.

The surprise or conflict is that the Porsche 944 wins on points while the panel preferred the Corvette, overall and with price as a consideration.

This isn't the contradiction it appears to be. Instead, it's emotion versus reason.

The Corvette is a live performance. The energy and enthusiasm come across the footlights with such irresistible power that the occasional clinker is forgiven.

The Porsche is a compact disc, without the raw appeal but because the recording studio preserves the sound so well, the gain makes up for the loss, or perhaps the two are enjoyed in different ways.

The RX-7 Turbo is in tune and in time, but it's done with synthesizers, those electronic devices that sound like violins or cellos or whatever . . . except they don't.

Mazda's engineers have done their lab work and their proving ground work. What they need now is to join the Corvette and Porsche engineers, out on the road in the real world.

CUMULATIVE RATINGS–SUBJECTIVE EVALUATIONS

	Chevrolet Corvette	Mazda RX-7 Turbo	Porsche 944 Turbo	Comments
Performance				
Engine	8	5	7	V-8 power and instant response beat top-end boost.
Gearbox	6	6	7	The Porsche gearshift never misses.
Steering	8	6	9	Porsche is precise; Corvette, vague; RX-7, overpowered.
Brakes	9	6	8	Corvette brakes can't lock, RX-7 rears do lock, Porsche lags.
Ride	6	4	7	Light and firm beats heavy and firm beats overcontrolled.
Handling	9	6	9	Two ways to do it right, versus one way to overdo it.
Body structure	6	6	8	The 944 is solid; the other two get the shakes.
Average	7.4	5.6	7.9	
Comfort/Controls:				
Driving position	8	8	8	Three sizes fit all, so we called it a draw.
Controls	7	7	8	Porsche controls all where you want; others are a hunt.
Instrumentation	7	7	8	Sometimes things look best in black and white.
Outward vision	6	7	7	Rear vision not quite as good on the Corvette.
Quietness	6	5	7	The silent payoff for the most solid construction.
Heat/vent/air cond	8	8	7	Porsche had less output.
Ingress/egress	5	7	6	The RX-7 has the largest openings, front and rear.
Front seat	7	8	8	Corvette's seat is over-adjustable.
Luggage & loading	4	7	6	RX-7 has the most space and the least reach for it.
Average	6.4	7.1	7.2	
Design/Styling:				
Exterior styling	9	7	8	Corvette is distinctive; Porsche, familiar; RX-7, derivative.
Exterior finish	7	8	7	Train the robots right, and they do it better than people.
Interior styling	6	7	8	Logical in the Porsche, fanciful in the other two.
Interior finish	8	8	8	Another draw and another set of high marks.
Average	7.5	7.5	7.8	
Overall average	7.0	6.7	7.6	
Staff preferences:				
Price independent	8	3	7	
Price dependent	8	3	7	

Ratings obtained through consensus after discussion.

Letter from Japan

BY JACK YAMAGUCHI

MAZDA RX-7 INFINI

Japanese designer sports car

Mazda's infinitely superior RX-7 Infini. Numbers aren't used on the Infini model so there is no RX-7 designation on the back end, just a Mobius ring.

PRONOUNCED "AN-FIN-KNEE," as the French do, the Infini Mazdas are special. The label was first applied to the twincam, 16-valve, turbocharged and intercooled 323. But the rapid hatchback was overshadowed by the 4wd 323.

The latest RX-7 Infini is a limited-production version of the RX-7 Turbo. There is no radical departure from the normal car; instead, it is a simplified version as far as equipment is concerned. The rear seat is eliminated for the first 2-seat RX-7 offered in Japan, and there isn't even a base AM radio in the package. The car has thus shed some weight, 88 lb compared with the Japanese GT-X version. Better seats are added, similar to those fitted in the pricier U.S. models, along with a 14½-in. leather-covered Momo steering wheel and leather-covered shift knob.

A straightforward turbocharged rotary engine, rated at 185 bhp JIS at 6500 rpm, is mated to either a type-R 5-speed or a 4-speed automatic. No change there.

What is significantly changed is the suspension. Wheels are now single-piece forged aluminum with wider 6½-in. rims, designed by BBS and produced by a BBS subsidiary in Japan just for this car. Each wheel is 5.5 lb lighter than the stock 6-in.-wide aluminum wheel. Tire size is unchanged at 205/60R-15. Additional weight is removed by fitting the U.S.-spec aluminum hood.

Mazda's computer-controlled adjustable shocks are not used on the Infini. Special shocks, tailored to the RX-7's suspension, are fitted. Two specialists submitted complete sets for Mazda's evaluation. The engineers found that one company had the best shocks for the front, while the other had the best choice for the rear, so that's what comes on the Infini. Damping is similar to the automatic suspension's firm setting; however, the transient characteristics have been smoothed with multi-spring discs in the valves. Piston diameters are larger and so is the outside diameter of the rear shocks.

The Infini handles and rides beautifully. The shocks are obviously more firm than the base RX-7's, but they work so smoothly and progressively that the ride hasn't suffered. And saving 22 lb of unsprung weight does wonders in allowing tires to follow surface irregularities. Power assisted steering is exactly as offered in the normal RX-7, but it feels more natural. The smaller-diameter Momo helps here, but I suspect they must have decreased the assist at low speeds. To use R&T's analogy, the feel of the car is no longer that of a synthesizer but is at least a well recorded compact disc.

Mazda has no plan to introduce the Infini in the U.S., though I told Takaharu Kobayakawa, who is in charge of the RX-7 program at Mazda, it would show the world that Japan can handcraft an outstanding car.

MAZDA RX-7 CONVERTIBLE

Perfect for a search-and-enjoy mission on a country road

PHOTOS BY BRIAN BLADES

Yes, Mazda is doing its share in the revival of the True Sports Car. The market for convertibles has almost tripled in the last five years in the United States, which made a convertible RX-7 a particularly tempting project. Whatever Mazda's reason—tradition, a humanitarian concern for the True Enthusiast, maybe even the profit motive—we should be grateful, because 6000 of the 35,000 RX-7s headed for the U.S. this year will be convertibles. The RX-7 open-air fare will be $20,500, and the only options are a $2400 package that adds leather-covered seats, cruise control and a compact disc player with stereo speakers in the headrests; and $859 air conditioning.

In making the RX-7 into a convertible, those fussy Mazda folks wanted more than an RX-7 without a top, explained product program manager Takaharu Kobayakawa. He said one of the biggest drawbacks of convertibles has been air turbulence. So, a goal was to offer top-down motoring joys, especially at highway speeds, without the occupants feeling like the cast of an Irwin Allen movie about hurricanes.

To accomplish this, Mazda engineers took aerodynamics into the passenger compartment. Their studies indicated that much of the wind buffeting the occupants of a convertible has passed over the windshield and then is ricochetted from the rear deck forward.

Enter the Windblocker (patent pending), a simple device consisting of a small panel located behind the rear seats. In its upright position (not even poking over the car's beltline), it's designed to deflect that reverse air flow.

Using the Windblocker does not eliminate buffeting, but there is a considerable and much-appreciated wind reduction, which also makes conversation easier. For those who like blowing in the wind, the Windblocker can be folded down, out of the way and out of service. But, there is more. The convertible is not just for summer anymore: An early morning sky is a terrible thing to waste and one tester found it extraordinarily pleasant to drive top down to work with the temperature in the 30s. This was possible with the heater on the lowest setting, the driver wearing a light jacket and a pretty pleased, aren't-I-special look. Mazda claims that, with a temperature of 50 degrees Fahrenheit and the RX-7 traveling at 50 mph, the temperature around the driver's right knee increases from 60 to 90 degrees, thanks to the Windblocker and a helpful heater.

Even winter use is possible, because the heater has such an easier time helping the RX-7's occupants avoid hypothermia. Imagine glancing out your window at work and exclaiming to coworkers: "Darn, it's snowing and I left the top down."

The aerodynamics outside aren't bad, either. With the top up, the Convertible has a coefficient of drag of 0.33, while with the top down, this increases only about 15 percent, to 0.38.

The top, which has a heated, glass rear window, is electrically actuated and remarkably easy to use. As on the Saab 900 convertible, the first step in its operation is safety-related: pull up the hand brake (Saab and Mazda recommending a stationary car, even if you are Rod Millen or Erik Carlsson).

Then, one unsnaps the retaining clips at the front, rotates a dash-mounted control and the top retracts. However, one must leave the car to release a lock folding the top's rigid front panel into place and to install the tonneau cover. Mazda officials say the rigid panel provides better weather sealing than a conventional design and indeed, during the Convertible's introduction an afternoon of extremely hard rain was thoughtfully arranged to demonstrate how well this works.

The top can also be used in what Mazda calls a half-top position, which is similar to the Targa arrangement favored by Porsche or the cabriolets of an earlier era altogether. The rigid front panel is simply unlatched and stowed in the trunk.

AT A GLANCE

	Mazda RX-7 Convertible	Alfa Romeo Graduate	Chevrolet Corvette Convertible
Price, base/ as tested	$20,500 $23,759	$15,400 $15,400	$34,820 est $37,830
Curb weight, lb	3030	2495	3320
Engine/drive	2-rotor/rwd	inline-4/rwd	V-8/rwd
Transmission	5-sp M	5-sp M	4-sp M + OD
0–60 mph, sec	9.7	10.4	6.3
Standing ¼ mi, sec @ mph	17.5 @ 81.0	17.6 @ 76.0	14.8 @ 95.0
Stopping distance from 60 mph, ft	145	152	130
Lateral acceleration, g	0.82	0.77	0.87
Slalom speed, mph	63.3	na	64.9
Fuel economy, mpg	16.0	23.8	est 17.0

	Pro	Con
RX-7 Convertible:	modern; good attention to top-down wind flow; very rigid chassis, good handling	only moderate acceleration, limited visibility with top up, modest luggage space
Graduate: tested 8-85	a classic sports car, timeless looks, easy to operate top, attractive price, charm	cowl shake, dated, rough engine; awkward driving position, modest handling limits
Corvette Convertible: tested 6-87	great looks, open-air fun, tremendous grip; good torquey acceleration; excellent ABS	heavy, price creeping up, limited luggage space, awkward ingress/egress

Like many convertibles, when the top is up the rearward vision suffers, with the B-pillar being the major offender. No matter how carefully the too-small side mirror is adjusted, there is still a hold-your-breath element to lane changes or, better yet, a brief askance glance.

As all God's children know, one of the problems with convertibles is the loss of structural rigidity. Mazda tried to compensate by using a combination of thicker steel around areas such as the rocker panels, A-pillars and the drivetrain tunnel. In addition, cross members were installed behind the engine, rear seats and between the rear struts.

The result is a convertible with about 70 percent of the torsional rigidity of the hardtop. To try to keep the weight down, the Convertible has an aluminum hood, just like the RX-7 Turbo. Nevertheless, the car we tested weighs about 240 lb more than a similarly equipped GXL coupe tested in February 1986.

But, happily, we found the fore aft weight distribution is still 50/50, just as on the GXL.

On the road, the body reinforcement works wonderfully

... those fancy Mazda folks wanted more than an RX-7 without a top ...

Top to bottom: Release windshield frame latches, electrically retract until half down (a 3-second pause allows clearing objects behind seats), manually fold targa panel, retract completely, cover with boot.

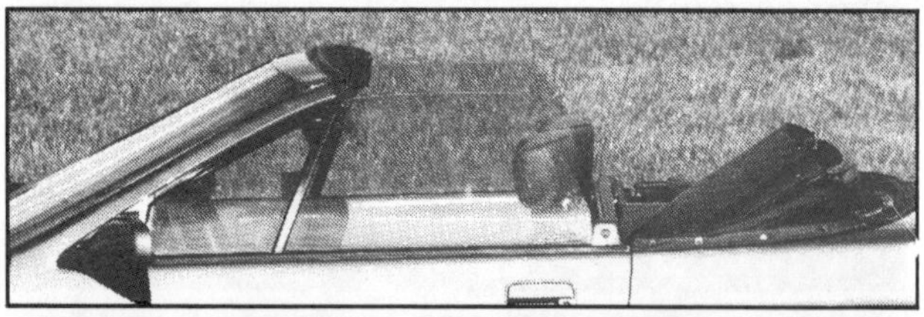

well. The Convertible feels solid, with virtually no indication that the roof has been removed. The exception is on particularly rough roads, the type normally found only in winter-battered regions. In such cases, there is some mild cowl shake, but the car is so generally delightful that it is a price worth paying. Mazda also gets excellent marks for a rattle-and-creak-free interior, regardless of how brutal the road surface might be.

The interior is virtually identical to that of the coupe, with the same generally thoughtful arrangement of driving controls, including the adjustable steering wheel and good heel-and-toe pedal arrangement.

The seats are form fitting, with good support, although adjustable wing bolsters would be a nice addition. Indeed, even folks of average girth noted just how form-fitting these seats are; you can imagine what our portly sorts think.

The Convertible does lose 0.5 in. of headroom (with the top up, smart alecks), but it can still easily accommodate a 6-ft driver. The instrumentation is mostly good and a little bad. The RX-7 has all the gauges a serious driver needs, with the tachometer being the largest and most prominent, placed directly in front of the driver. The bad news is that Mazda has stuck with its red-on-black color scheme, which is sometimes difficult to read. "Please, Mazda, rediscover white-on-black lettering, or something with contrast," one tester noted.

The only other complaints were that the stereo has enough buttons to baffle an astronaut and one person suggested Mazda should eliminate the fake stitching on the dash, even if it means the cruel firing of those fake seamstresses.

The amount of luggage space has been reduced from 19.1 cu ft in the 2-seater hatchback coupe to a "minimalist" 6.9 cu ft, according to our measurements. The spare tire, incidentally, is hidden behind the seats.

When it comes to power, the Convertible gets the normally aspirated twin-rotary 13B motor, which produces 146 bhp horsepower at 6500 rpm and 138 lb-ft of torque at 3500 rpm. However, one Mazda official suggested if there was enough demand for the 182-bhp turbochcharged rotary, that was a possibility in the future. The 13B motor is unchanged from the coupe's, with the same surprise-you-clod-you've-just-popped-the-redline-again smoothness.

For the speed freaks, there is a price to be paid in open-air motoring. The extra weight can be felt, particularly at lower engine speeds. "The bottom end is not great if you find yourself one gear higher than needed," a driver noted. "It's no V-8 . . . but it's smooth and sounds good and is generally delightful if you use the gearbox to best advantage."

The Convertible takes 9.7 seconds to reach 60 mph, 1.2 sec longer than the GXL tested in 1986. The slower acceleration is due not only to the additional structure, but also because Mazda has changed the final-drive ratio from 4.10:1 to 3.91:1 for better fuel economy.

A second here or there may not mean much when the top is down and all's right with the world. But, some RX-7 fans are likely to find the decreased acceleration a disappointment; it would be wonderful to see the turbocharged 13B offered.

We drove two different test cars, both equipped with 5-speed gearboxes that proved remarkably, annoyingly balky on chilly mornings. When fully warmed, they are more-than-adequate, but lack the easy, slick precision of the best linkages.

However, the gear ratios are well spaced. "You can do a lot of mountain driving just leaving it in 3rd, with useful power between 4000 and 7000 rpm," said a tester who took the high road. There is a considerable drop between 4th and 5th, making for quite relaxed highway cruising. But a quick pass on a 2-lane road will profit from a drop to 4th.

The all-independent suspension is standard RX-7, with some parts snitched from the Turbo to help cope with the additional weight.

ROAD TEST MAZDA RX-7 CONVERTIBLE

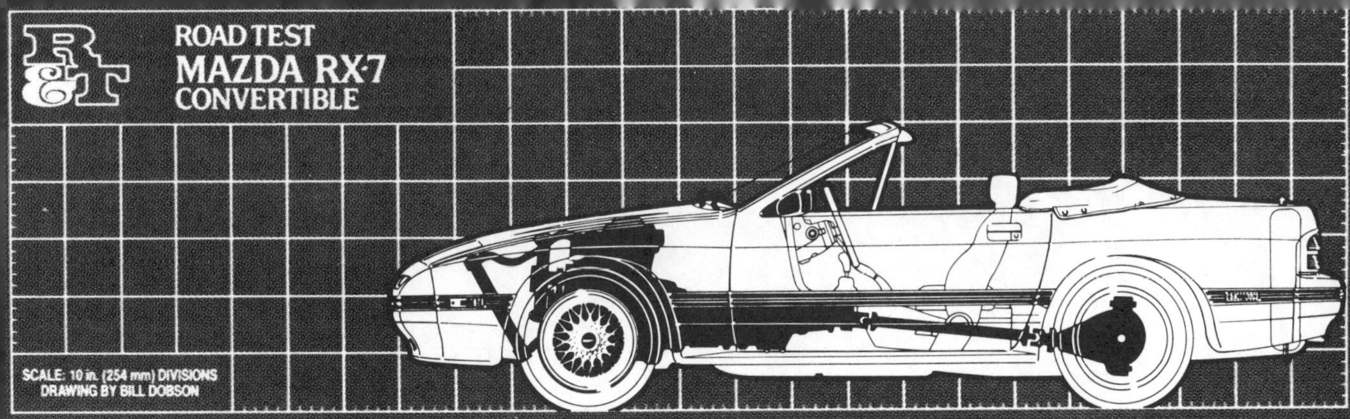

PRICE

List price, all POE $20,500
Price as tested $23,759
Price as tested includes std equip (elect. window lifts, elect. adj mirrors), option package (leather seats & steering wheel, AM/FM stereo/cassette/compact disc, headrest speakers, cruise control) $2400, air cond ($859).
Warranty: basic 36 mo/50,000 mi.

IMPORTER

Mazda Motors of America (Central), Inc, 7755 Irvine Center Dr, Irvine, Calif. 92718

GENERAL

Curb weight, lb 3030
Test weight 3170
Weight distribution (with driver), f/r, % 50/50
Wheelbase, in. 95.7
Track, f/r 57.1/56.7
Length 168.9
Width 66.5
Height 49.8
Ground clearance 4.8
Overhang, f/r 34.5/38.7
Trunk space, cu ft 6.9
Fuel capacity, U.S. gal. 16.6

ENGINE

Type .... 2-rotor Wankel
Bore width x chamber major axis, mm .... 80.0 x 240.0
Displacement, cc 1308
Compression ratio 9.4:1
Bhp @ rpm, SAE net .... 146 @ 6500
Torque @ rpm, lb-ft .... 138 @ 3500
Fuel delivery .... electronic injection
Fuel requirement .... unleaded, 87-pump oct
Exhaust-emission control equipment: 3-way catalytic converter, exhaust-gas recirculation, oxygen sensor

DRIVETRAIN

Transmission .... 5-sp manual
Gear ratios: 5th (0.71) 2.78:1
4th (1.00) 3.91:1
3rd (1.37) 5.36:1
2nd (2.00) 7.82:1
1st (3.48) 13.61:1
Final drive ratio 3.91:1

ACCOMMODATION

Seating capacity, persons 2
Head room, in. 36.0
Seat width .... 2 x 20.5
Seatback adjustment, deg 30

CHASSIS & BODY

Layout .... front engine/rear drive
Body/frame .... unit steel
Brake system, f/r .... 10.9-in. vented discs/10.7-in. vented discs, vacuum-assisted
Swept area, sq in. 397
Wheels .... alloy, 15 x 6½
Tires .... Bridgestone Potenza RE71, 205/60VR-15
Steering type .... rack & pinion, pwr assisted
Overall ratio 15.2:1
Turns, lock-to-lock 2.7
Turning circle, ft 32.2
Front Suspension: MacPherson struts, A-arms, coil springs over tube shocks, anti-roll bar
Rear Suspension: trailing arms with camber-control arms on drop links, semi-trailing links, floating hubs, coil springs, tube shocks, anti-roll bar

INSTRUMENTATION

Instruments: 150-mph speedometer, 8000-rpm tach, oil press., coolant temp, volts, fuel level

CALCULATED DATA

Lb/bhp (test weight) 21.7
Bhp/liter 111.6
Engine revs @ 60 mph in top gear 2200
R&T steering index 0.87
Brake swept area sq in./ton 251

ROAD TEST RESULTS

ACCELERATION

Time to distance, sec:
0–100 ft 3.4
0–500 ft 9.3
0–1320 ft (¼ mi) 17.5
Speed at end of ¼ mi, mph 81.0
Time to speed, sec:
0–30 mph 3.0
0–40 mph 5.1
0–50 mph 7.2
0–60 mph 9.7
0–70 mph 13.1
0–80 mph 17.0
0–90 mph 22.3

SPEEDS IN GEARS

Maximum engine rpm 7000
Rpm limiter .... none
5th gear (rpm) mph, est .... (4600) 125
4th, est .... (6600) 125
3rd 92
2nd 63
1st 36

FUEL ECONOMY

Normal driving, mpg 16.0

BRAKES

Minimum stopping distances, ft:
From 60 mph 145
From 80 mph 245
Control in panic stop .... very good
Pedal effort for 0.5g stop, lb 26
Fade: percent increase in pedal effort to maintain 0.5g deceleration in 6 stops from 60 mph .... nil
Overall brake rating .... very good

HANDLING

Lateral accel, 100-ft radius, g 0.82
Speed thru 700-ft slalom, mph 63.3

INTERIOR NOISE

Idle in neutral, dBA 49
Maximum, 1st gear 76
Constant 30 mph 63
50 mph 68
70 mph 73

ACCELERATION

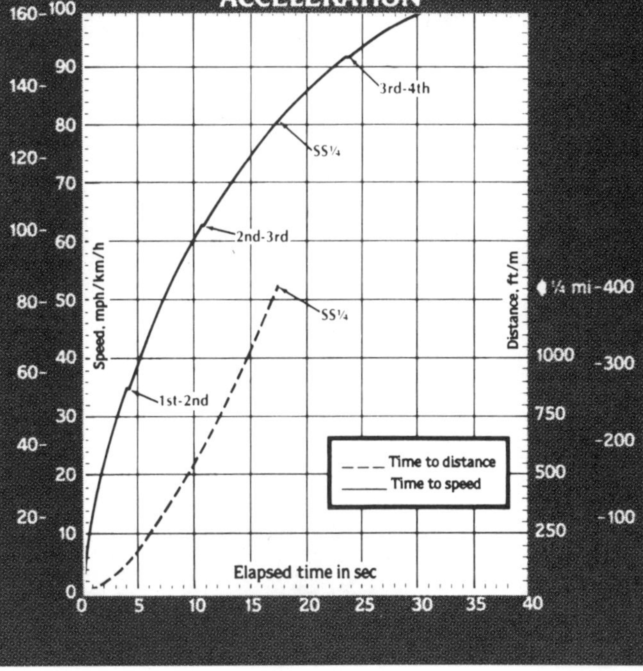

. . . *Mazda engineers took aerodynamics into the passenger compartment.*

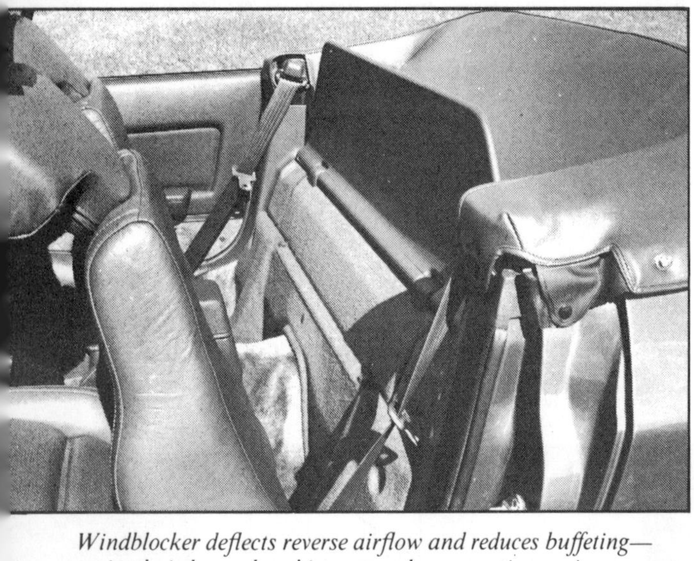

Windblocker deflects reverse airflow and reduces buffeting—saving hairdos and making normal conversation easier.

The Convertible also gets 15 x 6½ BBS wheels, with 205/60VR-15 Bridgestone RE71s. The wheels are slightly wider than the 15 x 6s found on the GTU and GXL models, but less aggressive than the 16 x 7 on the 55-series-tired Turbo. With the sky above and all that suspension below, the Convertible is perfect for a search-and-enjoy mission on a country road.

One quickly becomes extremely fond of the Convertible, with its satisfying agility and confidence-inspiring balance. Its basic nature is slight understeer. If you are interested in a little oversteer, you'll have to tromp the throttle pretty hard, lift in mid-corner at the appropriate time and pace or behave with considerable abandon on a wet surface.

"Nimble, but no Corvette in quick transitions," said one note writer, in itself high praise.

"It holds the road in a drift with good balance and just the slighest hint of understeer," observed another staffer. "You can really fling the car around, even on a wet road and it doesn't bite you."

The Convertible is not available with the Mazda's Auto Adjusting Suspension, but that feature was not particularly missed. Several people liked the predictability of the Convertible's normal suspension. "I'd rather drive and think for myself," said one.

The steering is rack and pinion, with a variable power assist. At low speeds, it can abruptly become too light. At higher speeds, the feel and effort are reasonably good, but it lacks the dialed-in sensation of the Porsche 944, for example. "There is a slight softness in the steering response that suggests more 300ZX than Vette here," one person observed.

A visit to the skidpad produced 0.82g, slightly less than the 0.84g we recorded in 1986 with a lighter-weight Sport Package RX-7 or 0.83g with a GXL. The Convertible's slalom speed was 63.3 mph, bracketed by 64.3 mph with the RX-7 Sport and 62.0 mph with the GXL.

The Convertible gets the same brakes as the Turbo, 10.9-in. ventilated discs up front with 4-piston aluminum calipers and 10.7-in. diameter ventilated discs in the rear. The brakes have good balance and a reassuring, linear response: How hard you push is how hard you stop.

Of course, it would be nice to have ABS, but the Convertible's stopping distances sans ABS were quite good. From 80 mph, it took just 245 ft to stop, a first-rate performance and 14 ft less than a 1986 Turbo.

In total, the RX-7 Convertible is a thoroughly delightful package, from the Windblocker eye-of-the-storm feature to Mazda's pleasing new twist on torsional rigidity. If you're having a mid-life crisis, this is a solution. If you've survived your mid-life crisis, you deserve one of these for your past suffering. And, if your mid-life crisis is still waiting in ambush, consider the RX-7 Convertible as 3030 lb of easy-to-swallow preventative medicine.

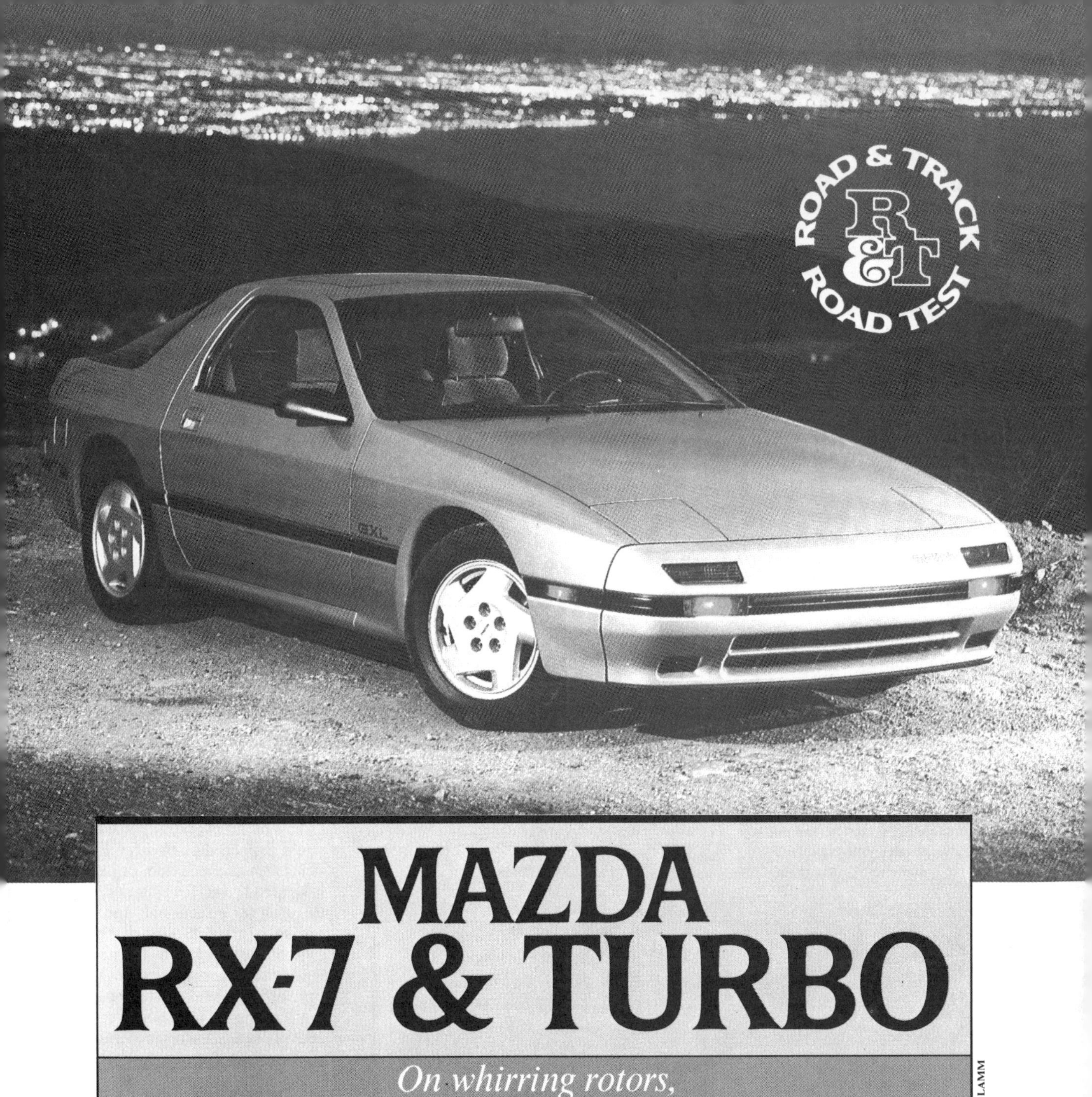

MAZDA RX-7 & TURBO

On whirring rotors, floating hubs and technology as its own reward

PHOTOS BY JOHN LAMM

MAZDA OF ALL THE world's sports-car makers, the one most committed to probing technology is Mazda. Too heady a statement? Consider: It was Mazda that brought the Wankel successfully to market, after Germany's NSU abandoned a considerable head start and even giant GM cut and ran. It was Mazda that kept with the rotary program, through initial sealing problems and later fuel economy pressures, to the point of the current intercooled turbocharged powerplant. And today it's Mazda with production features such as automatically adjustable and driver-selectable damping, electronically boosted steering and self-compensating floating-hub suspension, even as high-tech European automakers are openly deriding these concepts (while frantically working on their own responses).

What's more, all this technology is embodied in a single car, the Mazda RX-7.

Make that a single nameplate, the RX-7, because in fact Mazda has developed this sports car into a mini-line of several distinct models. At $14,199 the Standard RX-7 is powered by the same 13B engine as the rest of the line. At the other end of the normally aspirated extreme is the $18,449 GXL, complete with power assisted and automatically adjustable everything. A Luxury Package version ($15,799) fits in between there somewhere. And then there's the $20,399 Turbo. This one boasts not quite all the

GXL's automatic mods (no computer-controlled damping and steering, for instance), but also boasts boost. Not to be omitted, especially by us, for goodness' sake, is the Sport Package. At $14,749, this one is Mazda's lean and hungry RX-7, carrying nothing that doesn't make it go fast, stop quickly and corner like bejesus.

A quandary: Which to covet?

Each comes wrapped in bodywork that's aerodynamically efficient (an impressive 0.31 C_X for the Standard model; 0.29 with the Sport's skirts and spoiler), but not completely without controversy. "Derivative" sums up the RX-7's styling, with some evident Porsche influence here, a bit of Camaro/Firebird there, all adding up to a car that's handsome without being at all distinctive. "Kind of the opposite of the first-generation RX-7," was how one staff member put it. Then again, standing alone its shape is well integrated indeed, with nary a tack-on geegaw or break in its tautly developed surfaces.

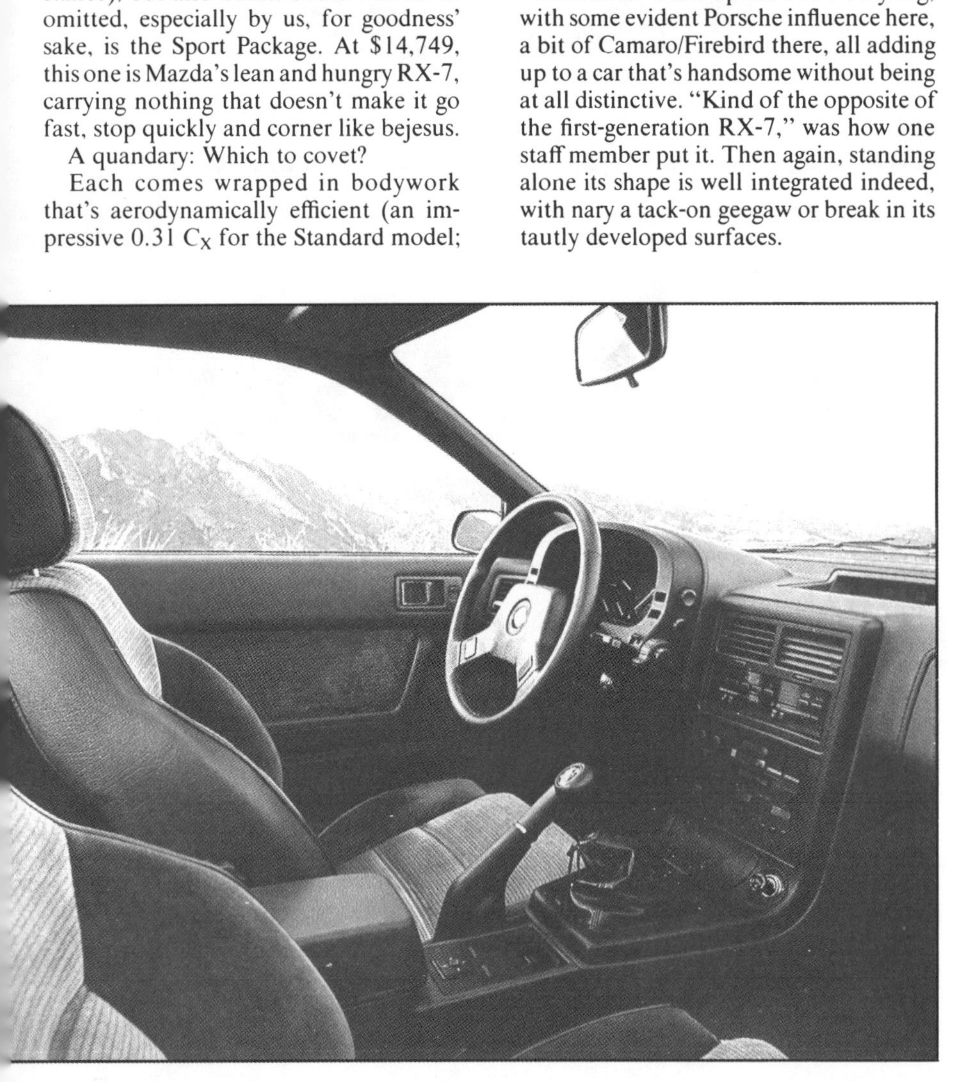

Hop behind the wheel of any model, and you'll be impressed by a well executed cockpit feel, perhaps just a tad on the jittery-jukebox-Japanese side of things. Gauges are all analog, thankfully, dominated by that oversize 8000-rpm tach. Some of us liked the red on black labeling; most found it difficult to see under all but best-case lighting.

You'll find everything just about where you'd expect, though a few controls may strike you as overstyled, or just plain silly. L-shaped levers, for instance, sprout from the dash for directionals and (if so fitted) cruise control. Some of the binnacle switches are masked by the steering wheel. And fake seamstresses spent far too much time on all the fake stitching. Detail carping (koi-ing?) aside, however, the RX-7's cockpit is a comfortable and functional environment for two enthusiasts.

But don't get your hopes up about the +2 option. Your dog would be just as happy in the back of the 2-seater.

The dog may not know a Wankel engine from a fire hydrant, but this is the characterizing feature of the RX-7. This twin-rotor powerplant has 6-port tuned induction (Dynamic Effect Intake is Mazda's moniker for it) and produces 146 bhp at an astonishingly smooth 6500 rpm. And if you've never driven a Wankel-powered car, you'll have to recalibrate your meaning of the word "smooth."

Propelling a full-house GXL, this engine gave us a 0–60-mph ride in 8.5 seconds and quarter-mile figures of 16.5 sec and 85.0 mph. This gets right in there and mixes it up really well with one of the RX-7's obvious competitors, the Porsche 944. Go play with your Mazda and your friend's Porsche, and you'll both be explaining it to the judge long before you've proved which one is quicker.

Ah, but you're going to get the Turbo, are you? This one has a dual-chamber turbo design, the point of which is to enhance low-end torque and minimize the time you wait around after punching the go pedal. Expect to find around 182 bhp whirring out of this engine at 6500 rpm. And even torque gets a 32-percent boost, from the 13B's normally aspirated 138 to 183 lb-ft at the same 3500 rpm.

Get this power to the road and you'll see 0–60 acceleration in the mid/high 6s. We hedge a bit because our first Turbo scooted there in 6.6 sec; a later one exhibited a bit less punch. It took around 7 sec. Certainly no slouch and, what's more, we sensed an extra dose of refinement that wasn't apparent in the original test car.

At any rate, we've faced off the RX-7 Turbo against the likes of Chevrolet's Corvette and Porsche's 944 Turbo. And none of the three had to blink.

Anti-lock braking is one of the few available options separating the 1987 RX-7 GXL or Turbo from their introductory-year brethren. At $1300 this is not your average check-and-forget add-on. But, believe us, ABS is worth it. On a clear, dry, straight surface, it'll make even a clubfoot as expert a braker as the R&T Shoe. His panic-stop figures with an earlier non-ABS RX-7 were superb: 133 ft from 60 mph and 243 from 80. His (and yours, for that matter) with our ABS Mazda work out to 133 and 239. And feel confident to challenge The Shoe when both your ABS car and his non-ABS are on any surface less than ideal. You'll beat him silly. What's more, ABS might even save a life someday. Not bad for $1300.

What about DTSS (Mazda's Dynamic Tracking Suspension System)? Mazda's semiactive, driver/car interactive suspension certainly places the RX-7 at the leading edge of technology, but not without some mixed reviews.

First off, what is it? DTSS is characterized by rear-wheel hubs that float on 3-point arrangements of bushings, each of the latter calibrated for varying compliance in different directions. As cornering forces increase, these hubs align themselves minutely from toe-out at low lateral g-force maneuvers through neutral to toe-in during really serious stuff. Additionally, control links maintain rear-camber angles at the optimum.

The idea is to enhance the balance between maneuverability at low and medium speeds and stability at the high end of the spectrum.

Now, on to the mixed reviews: The system evidently works, and very well indeed. How many cars do you know that turn 0.85g around a skidpad and 63.1 mph through our slalom? A glance through our data will convince you that these are world-class numbers. On the other hand, we were a little put off by a sense of artificiality, of it all being done with mirrors, not with sturdy bits of steel. We found the RX-7 a pleasant car to drive, a wonderful one to drive quickly, but a trifle disconcerting to drive at its limit. Or ours.

An earlier analysis put it in musical terms: "The Corvette is a live performance," we observed, "The Porsche is a compact disc, without the raw appeal but . . . the recording studio preserves the sound so well. The RX-7 Turbo is in tune and in time, but it's done with synthesizers . . ." There are those, no doubt, who want to get on right now with music of the future. Others are perfectly content with rebecs and sackbuts. We like to think of ourselves as somewhere in between.

So bring on the RX-7. We're ready for it. But can we arrange a drive of that MG TC too?—*Dennis Simanaitis*

Base price	$20,399
Price as tested	$21,699
Price as tested includes: std equip. (AM/FM stereo/cassette, limited-slip differential, 16-in. wheels & tires), ABS ($1300)	
GENERAL DATA	
Curb weight, lb	2830
Length, in.	168.9
Suspension, f/r	MacPherson struts/trailing arms with multi links, self-aligning
Brake system, f/r	10.9-in. vented discs/10.7-in. vented discs, ABS
Wheels	cast alloy, 16 x 7JJ
Tires	Goodyear Eagle VR, 205/55VR-16
Steering type	rack & pinion, variable power asst
Engine type	turbocharged 2-rotor Wankel
Displacement, cc	1308
Bhp @ rpm, SAE net	182 @ 6500
Torque @ rpm, lb-ft	183 @ 3500
Transmission	5-sp manual

PERFORMANCE	
Acceleration	
Time to distance, sec:	
0–1320 ft (¼ mi)	15.2
Speed at end of ¼ mi, mph	92.5
Time to speed, sec:	
0–30 mph	2.3
0–60 mph	6.6
0–80 mph	10.9
Braking	
Minimum stopping distances, ft:	
From 60 mph	133
From 80 mph	239
Control in panic stop	excellent
Handling	
Lateral accel, 100-ft radius, g	0.85
Speed thru 700-ft slalom, mph	63.1
Interior noise	
Maximum, 1st gear	75
Constant 70 mph	74
Fuel economy, mpg	19.2

MAZDA RX-7
THE FIRST DECADE

A test of time: 1979 RX-7, 1988 RX-7 GTU and 1988 RX-7 Turbo

PHOTOS BY RON PERRY

IT DOESN'T SEEM that long, but the RX-7 is 10 years old. To celebrate the occasion, Mazda laid three cars on us, one from the original 1978–1984 generation, two brand-new. The older car, a yellow 1979 model, was to remind us of a simpler, less expensive time when performance and driving pleasure were (almost) all that mattered. In other words, a sports car. The two new ones, a red GTU and a white 10th Anniversary turbo edition, were to emphasize the sophisticated technology and creature comforts that have moved the RX-7 (like almost all Japanese GT cars) decidedly upscale.

After Mazda's initial scare in the mid-Seventies—that the rotary engine could not live in the modern, fuel-efficient era—the company pulled the engine from its regular passenger lines and put it in an all-new sports car, one in which lively performance would more than overshadow its thirsty habits. The concept proved a resounding success and subsequent refinements to the rotary improved its fuel efficiency to a respectable if not truly thrifty level.

Our experience with the three cars confirmed these developments in no uncertain terms. With the second generation a lot has been gained, in performance, handling and luxury, while a little bit of the first generation's elemental directness, a no-frills honesty, has been lost, along with its down-to-earth price.

1979 RX-7

THE FIRST generation made a strong and lasting impression, and the 1979 coupe is an old friend. Our example had held up well, both physically and in its design elements, which are not all that dated after nine years.

The interior is roomy and comfortable, and the design of the instrument panel is still reasonably modern, with a strong grouping of dials in a cluster directly above the steering wheel, a tidy central console with ventilation and sound-system controls, and a gear lever that feels and works as well as the new ones. The engine is willing and the straight-line performance is not far below that of today's non-turbo GTU.

... the GTU is one of today's truly great driving machines, a superb balance of all the things that make a car go without sacrificing any of the qualities that make it nice to be in.

The 1979 interior is to modern standards of design and quality. With one Nikki carburetor, 1146-cc rotary puts out 100 bhp.

But before you conclude that we have observed no progress, we must mention the hollow slam of the thinner doors, the backfire of the thermal reactor emission control, a heavier steering effort, definitely less precision and grip on skidpad and in the slalom. All in all, it is a more basic, less refined machine in which all the operating systems are unmasked. For some, this is not criticism but praise; the entertainment comes immediately, at a lower performance level, and is sustained throughout. Proof that the original concept and most of the design details were right on the mark.

1988 RX-7 GTU

THE GTU initials refer to the IMSA racing category in which much of Mazda's performance image has been confirmed. Confirmed without doubt, with regular under-2.5-liter (that's what the U stands for) championships earned year after year. The production GTU is not really a spinoff of the racing machines (as is, say, a Ferrari GTO) but an honoring of the achievement. Compared to the 10th Anniversary turbo edition, it is a more basic version, with fewer frills and a more direct driver/road feeling.

The interior is still one of the best in a sports car. We say sports car, because its appearance and function are admirably direct, handsomely but not overdesigned, continuing the tradition of the 1978–1984 models. All the right gauges are right where they belong, starting with a large, prominent tachometer, and everything is quickly readable. As one of our staff put it: a haven of propriety in a world gone mad with digitals, video displays and beepers. The seats, Recaro-like, offer good support in all directions while remaining comfortable and unconfining for everyday use. And no silly power assist. The cargo bay of this 2-seater is roomy, perfect for two (but no more). As we said, it really *is* a sports car.

Despite revving to 3000 rpm on a cold start, the current rotary is more refined than that of the first generation, of course: quieter, smoother, more fuel efficient. It has a wide and usable if modest torque band, revs like crazy and dusts off almost any production car it comes up against. It *seems* a little slow off the mark—despite the added power, it is a heavier car than the original—but in fact it's marginally quicker. As with the original model, the shifter works positively, just a bit stiff when cold. →

GENERAL DATA

	1979 Mazda RX-7	1988 Mazda GTU/Turbo
Price		
Base price	$6995	$18,150/$24,650
Price as tested[1]	$7970	$20,100/$24,738
General		
Curb weight, lb	2420	2755/2900
Test weight	2515	2905/3050
Weight dist (with driver), f/r, %	53/47	51/49
Wheelbase, in.	95.3	95.7
Track, f/r	55.9/55.1	57.1/56.7
Length	168.7	168.9
Width	65.9	66.5
Height	49.6	49.8
Fuel capacity, U.S. gal.	14.5	16.6
Engine & Drivetrain		
Engine type	2-rotor Wankel	2-rotor Wankel
Chamber width x major asix, mm.	70.0 x 240.0	80.0 x 240.0
Displacement, cc	1146	130P
Compression ratio, :1	9.4	9.4/o.5
Bhp @ rpm, SAE net	100 @ 6000	146 @ 6500/ 182 @ 6500
Torque, lb-ft @ rpm,	105 @ 4000	138 @ 3500/ 183 @ 3500
Fuel delivery	one 4V carb	elect. fuel inj
Transmission	5-sp manual	5-sp manual
Gear ratios, :1, 1st	3.67	3.48/3.48
2nd	2.22	2.00/2.02
3rd	1.43	1.37/1.39
4th	1.00	1.00/1.00
5th	0.83	0.71/0.76
Final drive ratio, :1	3.91	4.10
Chassis & Body		
Layout	front engine/rwd	front engine/rwd
Body/frame	unit steel	unit steel
Steering type	recirculating ball	rack & pinion
Steering overall ratio, :1	variable, 17.0–20.0	20.3
Brake system, f/r	8.9-in. vented discs/ 7.9 x 1.3-in. drums, vacuum-assisted	10.9-in. vented discs/ 10.7-in. vented discs, vacuum-assisted
Wheels	alloy, 13 x 5½J	alloy, 15 x 6J/ alloy, 16 x7JJ
Tires	Bridgestone RD-106 185/70HR-13	Bridgestone Potenza RE71, 205/60VR-15/ Goodyear Eagle VR-55, 205/55VR-16
Suspension, f/r	MacPherson struts, lower lateral links, compliance struts, coil springs, tube shocks, anti-roll bar/live axle on lower trailing links, upper angled links and Watt linkage, coil springs, tube shocks, anti-roll bar	MacPherson struts, A-arms, coil springs, tube shocks, anti-roll bar/trailing arms with camber-control links, semi-trailing links floating hubs, coil springs, tube shocks, anti-roll bar

[1] Prices as tested include, for the 1979 RX-7 (1979 prices), air cond ($525), AM/FM stereo/cassette (est $200), alloy wheels ($250; for the RX-7 GTU, air cond ($859), elect. sunroof ($595), cruise control ($235), graphic equalizer ($149), floor mats ($55), center armrest ($53), std equip. (AM/FM stereo/cassette; for the RX-7 Turbo, glass-breakage detector ($88), std equip. (air cond, AM/FM stereo/cassette, elect. window lifts, elec. adj mirrors, central door locking, cruise control)

Handsome and logical, the GTU driver's office is a model of directness. The injected 1308-cc rotary develops 146 bhp.

The handling is sporting, with some harshness but less than you often get with this much grip. It still conveys its rear-drive configuration—hurrah!—but doesn't have the trailing-throttle oversteer that characterized the earlier RX-7s. The steering feels a little light, but that's a great improvement over the heaviness we noted in the 1979 car. Real progress. The brakes do the job well, but have a heavy feeling, and pedal effort isn't as easy to modulate as we'd like. It's interesting to note that the actual braking performances are remarkably similar in all three cars; the stopping distances are within a foot from 60 mph and just a few feet from 80!

All in all, the RX-7 GTU is one of today's truly great driving machines, a superb balance of all the things that make a car go without sacrificing any of the qualities that make it nice to be in. At $18,150 it isn't cheap, but it's certainly reasonable.

1988 RX-7 Turbo

MAZDA MAY have made a minor mistake in calling this turbo version the 10th Anniversary model, because its shift in purpose, away from the sports car and toward the overburdened luxury GT car, is a denial of the original concept. A denial but not a betrayal, because the unavoidable upward march of the GT market has also created excesses in the Nissan ZX and Toyota Supra, and the RX-7 compares more than favorably with them.

But for the way we like to drive, the Turbo isn't as satisfying as the GTU. The turbocharger gives more power, of course, but the lag is unpleasant and the top-end improvement isn't great enough—nor is it achieved often enough—to make it a

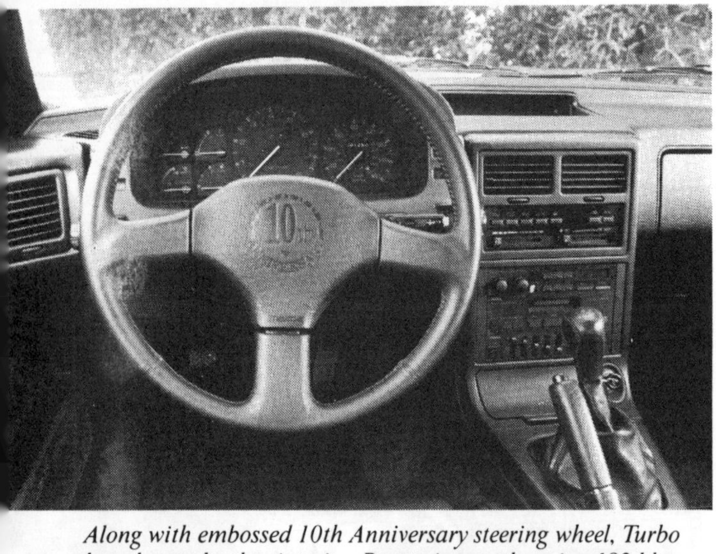

Along with embossed 10th Anniversary steering wheel, Turbo has elegant leather interior. Boost gives a whopping 182 bhp.

real improvement. And for some reason, Mazda hasn't bee able to get consistent, uniform quality in its turbo models. Ou test car, with only 1300 miles on it, had a clunky driveline emphasized by the poor low-end flexibility that made it hard t drive smoothly until one learned the trick of feathering th clutch and accelerator pedals. In fairness, the torque was bette than in our long-term Porsche 944, but this is damning praise

For an expensive GT car, the ride is rock-hard, with no com pliance or suppleness—again, like the 944. This must be a attempt to cope with the turbo power, but it is nevertheless lack of skill in tuning the chassis. And, considering that th model is the fully equipped luxury version for those who mus have everything, it is unnecessarily crude. Not capably raw i the RX-7 tradition, but rude.

It's too bad, because the 10th Anniversary RX-7 has som very nice things, such as the beautifully done leather interio with extremely comfortable seats and excellent ergonomic The concept of putting *everything* in the top model—powe grip, luxury equipment—without achieving an overall harmon is a mistake, for the ultimate satisfaction of the owner, eve though it can produce a strong showing in the marketplace.

The GTU is the *true* 10th-anniversary version of the RX-7 and a worthy one. We'll pick it any day.

PERFORMANCE

	1979 Mazda RX-7	1988 Mazda RX-7 GTU	1988 Mazda RX-7 Turbo
Acceleration			
Time to distance, sec			
0–100 ft	3.3	3.2	3.0
0–500 ft	9.2	9.0	8.0
0–1320 ft (¼ mi)	17.0	16.5	15.1
Speed at end of ¼ mi, mph	83.0	84.0	91.0
Time to speed, sec			
0–30 mph	3.0	2.7	2.2
0–60 mph	9.2	8.5	6.7
0–80 mph	15.8	14.9	11.2
0–100 mph	27.8	25.5	18.7
Est top speed, mph	122	128	140
Fuel economy, mpg	22.0	21.0	16.5
Braking			
Stopping distance, ft, from			
60 mph	151	152	150
80 mph	258	260	256
Pedal effort for 0.5g stop, lb	22	22	20
Brake fade (six 0.5g stops from 60 mph), % increase in lb	45	nil	10
Control	very good	good	very good
Overall brake rating	very good	very good	very good
Handling			
Lateral acceleration, g	0.78	0.80	0.83
Slalom speed, mph	59.3	64.0	64.6
Interior noise, dBA			
Idle in neutral	50	53	53
Maximum, 1st gear	77	73	71
Constant 30 mph	64	67	69
50 mph	68	71	69
70 mph	75	74	73
90 mph	78	na	75

CALCULATED DATA

	1979 Mazda RX-7	1988 Mazda RX-7 GTU	1988 Mazda RX-7 Turbo
Lb/bhp (test weight)	25.2	19.9	16.8
Bhp/liter	87.3	111.6	139.1
Engine rpm @ 60 mph in top gear	2850	2600	2800
R&T steering index	1.17	1.16	1.16

BY JACK YAMAGUCHI

MORE POWER TO THE JAPANESE

See the USA in your Mazda Cabriolet

PHOTOS BY THE AUTHOR

Mazda's RX-7 Cabriolet, the Japanese version of the U.S.-bound convertible.

OUR SPEED LIMIT is only 7 mph faster than that of the U.S., and highways and roads are generally overcrowded and narrow. By official decree, Japanese performance automobiles are equipped with speed limiters, usually cutting off fuel at 180 km/h (113 mph), as well as those insistent warning buzzers, beepers or bells that start sounding at 100 km/h (62 mph).

Yet, the Japanese want more power. The new Mazda RX-7 Cabriolet and the revamped Honda Civic are two such examples. The quick-change Mazda is powered by the turbocharged, intercooled version of the type 13B twin-rotor Wankel, which is the only engine choice in Japan. We may be highly regulated, but we haven't come to a Corporate Average Fuel Economy or Gas Guzzler Tax yet. The turbocharged 13B puts out 185 bhp JIS, which is not much different from the SAE net rating, compared with the naturally aspirated U.S. version's 146 bhp SAE net. Either model weighs 3000 lb, take or add a few, which is about 200 lb heavier than a comparably equipped coupe version.

The Japanese Cabriolet may be had with either 5-speed manual or 4-speed automatic transmission, the latter unavailable in the U.S. to avoid shoving the car into the dreaded Gas Guzzler Tax category. What the U.S. buyer gets, and we don't, is a set of those beautiful, lightweight BBS forged alloy wheels with 6.5-in. rim width. The Japanese version is fitted with 6-in. cast-aluminum ones.

The RX-7 convertible is equipped with a rather clever and simple device dryly called a wind-deflection board for America, but it carries the fancier name of aero-board for Japan. Regardless of nomenclature, the board is a brainchild of Takaharu "Kobby" Kobayakawa, chief engineer of the RX-7 range. Kobby is a skiing-freak and loves driving through the mountains in winter. Prior to his RX-7 assignment, Kobby was in charge of Mazda's product evaluation. When the original 323 Cabriolet hit the Japanese market, he insisted that the car should be tried in some adverse climatic conditions and took it to his favorite ski site near Hiroshima, with the top down! The hapless troupe got frozen, not so much from the front but by reverse flow. Kobby hastily devised a curtain that largely blocked this reverse flow. An optional aero-curtain is offered in the Japanese-version Mazda 323 Cabriolet, and for the new RX-7 open-top version, a sturdier and more effective folding wind-deflection board is installed as standard equipment.

Mazda MX-5 Miata

GRAND PRIZE

The lightweight sports car is back! And it's better than ever.

BY DENNIS SIMANAITIS
PHOTOS BY JOHN LAMM

IT WAS ALL very hush-hush at the time, but back last summer I was invited to Mazda's Myoshi proving grounds to drive a prototype of the company's all-new lightweight sports car. After a video briefing, artful in evoking the essence of sports car enthusiasm, I was encouraged to crawl around, under and into three of the prototypes on display. Then I got to drive the yet-unnamed lightweight sports car around Myoshi's Global Road Circuit. It was an absolute delight and my post-drive debriefing with Mazda engineers probably sounded like over-hyped ad copy.

I returned from Japan with sky-high enthusiasm for the car—and a pledge of strict secrecy.

But now the car has made its official debut at the Chicago auto show, so my pledge has been fulfilled, and I'm here to tell you all about it.

It's called the Mazda MX-5 Miata. MX-5 fits the car numerically into the company's lineup, in spirit with the RX-7 and MX-6, but more elemental. And the word Miata, Old Germanic meaning "prize," also has something of a Japanese ring to it that's not a bad idea at all, what with the car's origins and the good vibes these days concerning Japanese quality, durability and value.

The Miata is precisely an updated version of the classic lightweight sports car, a 2-seat convertible with front engine and rear drive. A latter-day Lotus Elan with styling that's simultaneously fresh and modern, yet recalling sports/racing cars of the Fifties. And it's every bit as much fun as I remember from last summer.

I had good opportunity to chat several times with Mazda's Toshihiko Hirai, product program manager for Miata development. And from these conversations, I learned a lot about the philosophical nuances of a lightweight sports car, as perceived by Mazda. As Hirai expressed it, for instance, the driver and car should form an integrated whole, like the oneness between rider and horse.

He shared a compilation of objectives defin-

ing what he hoped the lightweight sports car would be. I found the list especially enlightening on two counts. First, it provoked thoughtful objectivity about something that's largely a matter of the heart—enthusiasm for a mechanical contrivance. Second, it helped assess how brilliantly conceived and executed this new car is.

Its convertible top is a good example. It had to be easily erected or stowed from the driver seat. It required a latching mechanism that was easy to use. Erected, it had to offer perfect sealing up to 80 mph. Down, it had to stow completely below the car's rear deck line.

These engineering criteria give a perfect description of the Miata's soft top, developed in conjunction with the British design specialist IAD. Its latches, for instance, are large and of a novel over-center sort that doesn't need Godzilla for actuation. Though not a heavily padded top of the Germanic idiom, Mazda does get its top materials from the same firm supplying Volkswagen and BMW.

Like any proper sports car, the Miata looks better with its top stowed. But its top-up appearance has a certain jauntiness as well. And there's a hardtop too, unlined and elemental, but just the thing for snow-belt inclemency.

A proper powerplant for a proper sports car: a 1.6-liter inline-4, double overhead camshafts actuating four valves per cylinder, 116 bhp at 6500 rpm—all dedicated to driving the rear wheels.

With either top erected, the Miata measures 48.2 in. high, 1.0 in. lower than an MGB, for example. Continuing this not-inappropriate comparison, the Miata is a tad longer (155.4 versus the B's 153.1 in.) on a slightly shorter wheelbase (89.2 vs 90.9 in.) and a lot wider (65.9 vs 59.8 in.). Personally, I was surprised that the Miata is about MGB size; in real life, it looks smaller.

There are some styling cues echoing its bigger brother RX-7; around the windshield, for instance, and the hood and front fenderlines. But it also reminds me of a more rounded, more voluptuous rendering of a Lotus Elan. I especially like the front end, with that low air intake and a bumper that's so well integrated you don't even know it's there.

The wheels might remind you of Minilites from the Sixties (assuming you're of an age to remember *anything* from that decade). The tires, 185/60R-14s, are either Bridgestones or Dunlops, specially designed for the Miata.

Slip into the driver's seat and you'll find an interior in keeping with the Miata's spirit; simple, elemental, but by no means stark. It's less "styled" than the RX-7's interior. But there's lots of thought about design, feel and location. The sun visors, for example, fold in half along the windshield header so they don't obstruct vision when not needed. Handsome round gauges (analog, naturally) nestle in a hooded cluster immediately ahead. Controls for heat/vent/(optional) air conditioning and (standard

equipment) radio/cassette player reside in the center console.

Though diminutive, it's an airy cockpit meeting another of Mazda's goals: having just the right amount of turbulence enhancing the open-air experience, a direct contact with nature. There's a bit of room behind the seats for stowage, even when the top is down. The trunk adds space for a couple of soft bags.

Twist the key and you're rewarded with a pleasant burble from the Miata's 1597-cc fuel-injected inline-4, essentially the same dohc 4-valver powering sporty 323s. One change, a lightened flywheel, emphasizes its sports-car intent. Mr Hirai's wish list is evident in other ways as well, for several of its entries address engine sounds and noise (which, as any enthusiast knows, are two distinctly different things). A couple of these goals are familiar: a powerful low-frequency sound, and elimination of rpm-sensitized resonance or boom. Two others, though, were new to me, but spot-on in characterizing a pleasant engine: sounds that vary linearly with engine rpm, and also with throttle activity.

Snick the shift lever into 1st, and you'll experience another element on Mr Hirai's list. The shifter and its actuation are super short and wonderfully positive. It reminded me of

■ How to be elemental without being embarrassingly basic: Well-conceived details of the Miata emphasize litheness and function; and the best view is an action one.

■ Something old and something new, something styled and something blue. Does the Miata remind you of a Lotus Elan? An Alfa Romeo Duetto? An RX-7? Each of these, maybe, but refreshing too.

the best of Formula Ford race cars. Run up through the close-ratio gears in a reasonably aggressive way, and you'll probably reach 60 mph in something less than 8 seconds; adequately quick these days, and fine for good clean fun. Even better, torque is strong across a wide band, and I for one prefer this to the peakiness of some turbocharged powerplants that come to mind.

Consulting Mr Hirai's list again, engineers worked for a feeling of directness between throttle and rear wheels. I believe a lot of this driveline integrity is achieved by an artfully cast piece of aluminum that connects the engine, transmission and final drive. Check off another goal successfully met.

Handling goals occupy a good portion of the list, and it's in this area that the Miata impresses me the most. Its combination of communication, responsiveness, predictability and forgiveness makes it the best-handling 2-seater I've driven in recent memory—and my memory for such things is good.

The suspension has unequal-length A-arms, concentric spring/damper units and anti-roll bars, front and rear. Steering is by rack and pinion, with no real need I could sense for the power assist that'll be optionally available. Wish-list goals include things like quick and rigid response in steering, minimal camber change in cornering, passive toe-in at the rear and lateral rigidity of suspension components, each contributing to enhanced stability and grip.

Two other attributes, however, make the Miata suspension my current favorite. First, it's not afraid to transmit a certain amount of road irregularities, even some steering kickback, in telling you what's happening down at those tire contact patches. Thus, you're given particularly good information about grip. And, second, it responds crisply yet predictably to varying throttle angle. So there's just enough drop-throttle tuck-in, for example, to reposition the car without having to invoke increased understeer through added steering lock.

The brakes are discs all around, vented at the front. In keeping with the Miata's elemental nature, no ABS is offered. On the other hand, Mr Hirai's list includes things like a taut feel of the brake pedal, optimal fore/aft balance, amenability to modulation and easy facility for heel-and-toe. As is evident from this list, you know Mr Hirai is one of us.

Other genuine enthusiasts hang out at Mazda's North American R&D facility, not far from the R&T offices. Tsutomu (also known as a Californian named Tom) Matano, chief designer, advanced product development division, played a key role in the Miata's development. Mark Jordan, assistant chief designer, defined the car's handsome contours (and I suspect his father, Chuck, who does something or other at GM, could be more than a trifle envious, not to say proud). Bob Hall, an automotive journalist before going straight, gave the project the proper impetus in his product development role.

Before long, folks around the country will be able to drive Miatas of their own, due to reach eager buyers by late spring. Japan won't see any until September; Australians will have to wait until October. It'll arrive initially with precisely three choices of color: red, white or blue. They're especially attractive pure hues with nary a trace of metallic flake. Mr Hirai said there's some thought of test marketing two others, a bright yellow and British Racing Green.

Mazda hopes to find homes for some 40,000 Miatas a year, which doesn't sound all that difficult. Price, of course, depends on the dollar/yen relationship of the moment, but figure the logic of the situation: The Miata is worth more than the sportiest 323, but should come in at less than the least expensive RX-7. Sounds like $14,000–$16,000 to me.

Sounds like fun, too, because I surely remember last summer. Myoshi's Global Road Circuit, as its name implies, gives a quick around-the-world motoring tour. It has everything from my hometown Santa Ana's Interstate 5, complete with carefully hand-crafted freeway hop, to twisty Bavarian backroads missing only an Alp here and there. I was given the prototype Miata to do as many laps as I liked, at whatever pace I cared to set.

After a while, my hosts figured I would eventually just run out of fuel.

Wait till you drive the Miata! You'll love it.

SPECIFICATIONS

Curb weight	2180 lb
Wheelbase	89.2 in.
Track, f/r	55.5 in./56.2 in.
Length	155.4 in.
Width	65.9 in.
Height	48.2 in.
Fuel capacity	11.9 gal.

ENGINE & DRIVETRAIN

Type	4-valve dohc **inline-4**
Bore x stroke	78.0 x 83.6 mm
Displacement	**1597 cc**
Compression ratio	9.4:1
Horsepower (SAE net)	**116 bhp @ 6500 rpm**
Torque	**100 lb-ft @ 5500 rpm**
Fuel delivery	electronic fuel injection
Transmission	**5-sp manual**

CHASSIS & BODY

Layout	**front engine/rear drive**
Brake system, f/r	**vented disc/disc**
Wheels	cast alloy, 14 x 6
Tires	**185/60R-14**
Steering type	**rack & pinion**
Suspension, f/r	**unequal-length A-arms**, coil springs, tube shocks, anti-roll bar/**angled unequal-length A-arms**, coil springs, tube shocks, anti-roll bar

MAZDA MX-5 MIATA

What we're doing on our summer vacation

人馬一体

■ **These Japanese symbols (translation: "oneness between horse and rider") represent Product Program Manager Toshihiko Hirai's philosophy that was carried through the Miata's development.**

Can the Mazda MX-5 Miata turn water into wine? No, but it *has*, by and large, turned automotive journalists into jelly. Can it part the Red Sea? No, but it can pave a road across the Pacific Ocean to deliver motoring enthusiasts out of the desert and into the Promised Land of Sports Cars.

Can it update the dream of the inexpensive, lightweight, 2-seat convertible? Yes. Can it predict the future? Let's hope so.

More than any other car in recent memory—and even fairly distant memory—the Mazda Miata has created a stir of almost embarrassing proportions.

What could turn an otherwise jaded group of PC-wielding car writers into a bunch of gushing quill pushers, all romance and flowers? What is it about this car that makes CRX owners (among others) jump out of their Hondas at stoplights to inquire, "What is it? Who makes it? How much?" Or, further, to have an in-the-know driver follow you for blocks just to tell you gleefully, "I've got one on order!"

The answer, of course, is fun. Pure, simple, undistilled, clear-eyed driving fun; from the top of your head (a convertible lid) to the tips of your toes (a drilled accelerator pedal).

It was during the middle of this decade that some of Mazda's young Turks decided that what this world needed was a fun, bargain-priced, honest, classic—etc., etc.—sports car.

PHOTOS BY HENRY THOMAS

ROAD & TRACK
R&T
ROAD TEST

Today, the quality of the idea seems as obvious as an elephant in your living room, especially when the 1989 denizens of the sports-car fun zone are either gone (Pontiac Fiero, Bertone X1/9), expensive (the next-generation Toyota MR2 and Mazda's own RX-7) or front-drive (Honda CRX Si). And the only classic roadster of the bunch, the Alfa Romeo Spider, was 10 years old *10 years ago.*

At the time the Miata idea was proposed, however, Mazda felt that it was also about as practical as having said elephant in said living room. So the fathers of the MX-5 Miata held their breath during the car's tenuous five-year incubation, relying on faith and their own determination to keep the project going during the many times when the Miata's future looked bleak. As Toshihiko Hirai—leader of the Miata project and an older, wiser young Turk—expressed it, "There were times when a small breeze from the wrong direction could have blown the project away."

Mazda thought it was getting a niche car: a $14,000, front-engine/rear-drive, open 2-seater. What the Japanese automaker has in fact tapped into, however, is the recognition of a longing in the U.S. for inexpensive, lovable roadsters; a thirst that has gone unslaked for a decade. Factor in the consistency and reliability of Japanese manufacturing, and it is easy to see why the Miata is being treated like an avatar instead of an automobile.

But we tested it as an automobile. And as such, we discovered that the Miata is a great car with good, but not great, performance. We also found that while numbers don't lie, the whole of a car is often greater than the sum of its sums.

At 2205 lb., the Miata is about 100 lb. heavier than a CRX Si and 150 lighter than the current MR2—two good cars for comparison, because they, like the Miata, are powered by 1.6-liter, 4-valve-per-cylinder engines. No matter, both the Toyota and the Honda are significantly quicker off the line (0–60 mph) and back onto the dime (braking from 60 and 80 mph) than the Mazda.

The Miata's fuel-injected 116-bhp dohc four (the only engine Mazda offers) keeps giving all the way up to its 7000-rpm redline. Musically, the engine is a perfect tenor, serenading the driver lustily to 6200 rpm, where the exhaust note begins to sour.

This engine is mated exclusively (no automatic transmission offered) to a terrific 5-speed manual with super-short throws and a precise feel.

Fairly short gearing in 1st and 2nd gears gives the Miata driver the sensation of being able to launch the car quickly when the light turns green. The car's P185/60HR-14 Bridgestones also contribute to this feeling of initial quickness as they allow some off-the-line wheelspin in 1st and a tiny chirp in 2nd.

■ Simplicity without austerity characterizes the Miata's interior. Instrumentation is complete; large chrome-bezeled tach keeps tabs on the free-revving 1.6-liter 16-valve engine.

MAZDA MX-5 MIATA

PRICE

List price, all POE	est $13,800
Price as tested	est $15,550

Price as tested includes option package A (power-assist steering, AM/FM stereo/cassette, alloy wheels, leather steering wheel) est $1000, air cond (est $750).

IMPORTER

Mazda Motors of America, Inc,
PO Box 19734, Irvine, Calif. 92713

0–60 mph	9.5 sec
0–¼ mi	17.0 sec
Top speed	est 117 mph
Skidpad	0.83g
Slalom	62.4 mph
Brake rating	good

DRAWING BY BILL DOBSON

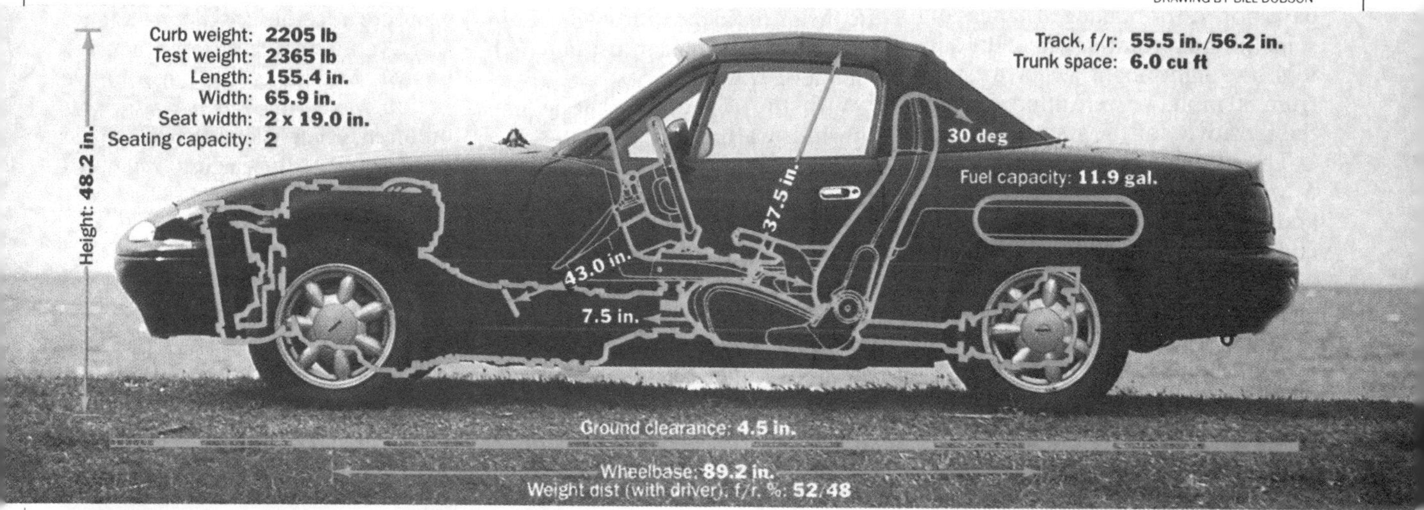

ENGINE

Type	iron block, alloy head, **inline-4**
Valvetrain	dohc, 4-valve/cyl
Displacement	97 cu in./1597 cc
Bore x stroke	3.07 x 3.29 in./ 78.0 x 83.6 mm
Compression ratio	9.4:1
Horsepower (net)	**116 bhp @ 6500 rpm**
Bhp/liter	72.6
Torque	**100 lb-ft @ 5500 rpm**
Maximum engine speed	7000 rpm
Fuel injection	Nippondenso elect. port
Fuel requirement	prem unleaded, 91 pump oct

CHASSIS & BODY

Layout	**front engine/rear drive**
Body/frame	unit steel
Brakes	
Front	**9.3-in. vented discs**
Rear	**9.1-in. discs**
Assist type	vacuum
Total swept area	285 sq in.
Swept area/ton	240 sq in.
Wheels	cast alloy, **14 x 5½JJ**
Tires	Bridgestone SF-325, **P185/60HR-14**
Steering	**rack & pinion,** pwr assist
Overall ratio	15.1:1
Turns, lock to lock	2.8
Turning circle	30.0 ft
Suspension	
Front	**upper & lower A-arms,** coil springs, tube shocks, anti-roll bar
Rear	**upper & lower A-arms,** coil springs, tube shocks, anti-roll bar

DRIVETRAIN

Transmission .... **5-sp manual**

Gear	Ratio	Overall ratio	(Rpm) Mph
1st	3.14:1	13.48:1	33
2nd	1.89:1	8.12:1	54
3rd	1.33:1	5.72:1	77
4th	1.00:1	4.30:1	102
5th	0.81:1	3.50:1	est (6510) 117

Final drive ratio 4.30:1
Engine rpm @ 60 mph in 5th 3340

INTERIOR NOISE

Idle in neutral	52 dBA
Maximum in 1st gear	84 dBA
Constant 50 mph	72 dBA
70 mph	77 dBA

FUEL ECONOMY

Normal driving	25.5 mpg
EPA city/highway	est 25/30 mpg
Cruise range	280 miles
Fuel capacity	11.9 gal.

INSTRUMENTATION

140-mph speedometer, 8000-rpm tach, oil press., coolant temp, fuel level

MAINTENANCE

Oil/filter change	7500/7500 mi
Tuneup	30,000 mi
Basic warranty	36 mo/50,000 mi

Test Notes . . .

■ Coupling the Miata's engine and differential is a light, but rigid, CAD-engineered truss that improves throttle response and reduces drivetrain shudder.

■ The Miata is a sports car full of subtle virtues that may need miles to be appreciated. Examples: The seats adapt to drivers of nearly any size, there's almost no cowl shake, the engine is easily accessible and you get an extraordinary ride for a short wheelbase.

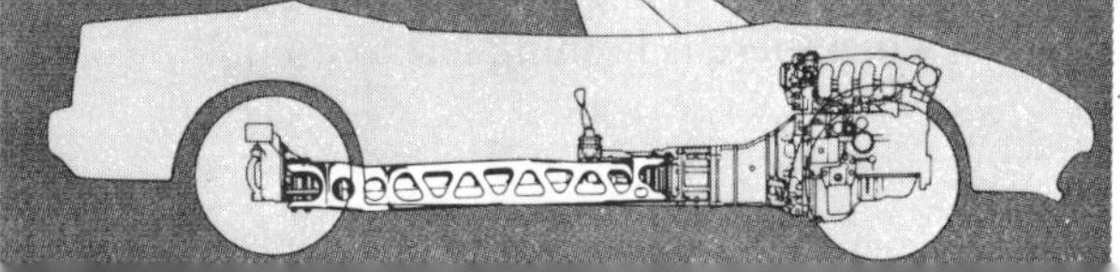

ACCELERATION

Time to speed	Seconds
0–30 mph	3.0
0–40 mph	4.8
0–50 mph	7.0
0–60 mph	9.5
0–70 mph	12.3
0–80 mph	16.5
0–90 mph	21.6
Time to distance	
0–100 ft	3.4
0–500 ft	9.2
0–1320 ft (¼ mi)	17.0 @ 81.5 mph

BRAKING

Minimum stopping distance	
From 60 mph	154 ft
From 80 mph	270 ft
Control	very good
Pedal effort for 0.5g stop	16 lb
Fade, effort after six 0.5g stops from 60 mph	22 lb
Brake feel	excellent
Overall brake rating	good

HANDLING

Lateral accel (200-ft skidpad)	0.83g
Balance	moderate understeer
Speed thru 700-ft slalom	62.4 mph
Balance	mild oversteer
Lateral seat support	very good

Subjective ratings consist of excellent, very good, good, average, poor.

But none of this changes the stock MX-5 Miata's best 0–60-mph sprint, which, at 9.5 seconds, is at least a full second slower than that of either the MR2 or the CRX Si.

The Miata's 4-wheel disc brakes don't seem to give it any relative stopping advantage, either. The brake feel is very good, however, with excellent communication between the pedal, brakes and tires. And while we would have hoped for braking distances of something shorter than the 270 ft. it took the lightweight Miata to stop from 80 mph, we were impressed by the sensitivity of the system.

The suspension setup on the MX-5 Miata is simple and direct, front and rear: upper and lower A-arms, coil springs over tube shocks, and an anti-roll bar. Mazda has set up the stock Miata for comfort more than handling. The ride is comfortably soft with more body roll than we would have expected.

All of this is not to say that the Miata, with its light weight and short wheelbase, isn't quick and responsive; it's both. But it could be quicker and even more responsive. Already it is easy to see the aftermarket explosion this car is going to cause. A tighter suspension and wider tires here, a few engine modifications there, and you could have the premier autocrosser of the Nineties.

So much for the stuff of engineers, now let's go to the poet's camp. We'll leave the styling analysis to the expert (see sidebar), but we will linger on the Miata's exterior long enough to appreciate the refreshment of a car body that looks as if it were grown from nature, not carved in a wind tunnel. Brazenly, the Miata flaunts its high coefficient of drag in the face of dozens of low-C_X clone cars.

That natural theme is carried through to the inside of the Miata as well. There are very few sharp edges to cut the eyesight on and no complication to clutter the mind (and precious little trunk space to accumulate any clutter *in*). The controls are laid out simply and directly, as are the chrome-bezeled gauges on the instrument panel.

The driver also fits into this environment as a natural consequence. The seats are comfortable for distance driving and supportive during the twisting fun runs. While just a tad shorter on leg room than an MR2, the Miata will still fit a 6-ft. 4-in. editor as if he were born to the car.

We even had the convertible top up one time to check on the head room (we found plenty). The rest of the time it was top-down traveling along the Pacific Coast Highway. The one morning when we did get a bit of rain, it was a simple matter to manually raise the top, zip up the plastic rear window and clamp the two latches into their seats in the windshield header. The top's sealing against rain and wind noise was superb.

With the top down, tne swirling winds wafted freely through the driving compartment, but did so without buffeting the occupants to distraction or noisily ruining conversation. It is also important to point out that chassis flexing, a crucial consideration (and problem spot) on convertibles, was virtually nonexistent in the Miata. Driver ego flexing, on the other hand, was rampant.

It is comforting to note that as a result of Mazda's determination to keep it simple and inexpensive, the MX-5 Miata is an easy car to keep track of in your mind. As noted earlier, there is only one engine/gearbox combination. Three colors are available—red, white and blue—with a hot yellow (as seen on the Club Racer version at the Chicago auto show) and British Racing Green (if you really want to rub it in) being considered for the future.

If you must have your sports-car cake and eat it too, two optional upgrade packages are available: Package A includes alloy wheels, an AM/FM stereo/cassette system, leather-wrapped steering wheel and power-assisted steering. Package B includes all the add-ons of package A plus the sins of cruise control, headrest speakers and—gasp—electric window lifts. Additionally, there are four straight-ahead options: a limited-slip differential, a compact disc player, air conditioning, and a detachable hardtop (only available in red as of this printing).

This road test of the Mazda MX-5 Miata is not a test of numbers. The objective performance results are here. You can look at them, mull them over, recalculate them for us, and even commit them to memory for future use and comparison. There will come a time, however, when there is nothing left to do but drive the car. Some sunny day, you'll put on your driving gloves, take the top down and head out to your favorite country road. You'll run through the gears, upshifting and downshifting, as one with the road, feeling out your favorite corner as though it were your first time.

At that moment, all of those numbers will just get blown away in the rushing wind like so many useless gasoline receipts in the passenger seat. And the smile on your face will be real. And the twinkle in your eye as you wave a gloved hand to your brethren, other Miata drivers, will be earnest and enthusiastic.

Mazda MX-5 Miata Styling Analysis

BY MARK STEHRENBERGER

RAME AROUND WINDSHIELD SPELLS SAFETY, BUT LOOKS TOO HEAVY FOR SIZE OF CAR. PARTIAL BLACKENING-OUT WOULD HELP.

FRONT PORTION OF CAR LOOKS SOMEWHAT TOO BULBOUS AND HEAVY. PROPOSE FLATTER HOOD LINE TO MAKE CAR LOOK LOWER, WIDER AND MORE AGILE.

ADDITION OF A LOGO OR BADGE TO NOSE WOULD BREAK UP LARGE EXPANSE OF PLASTIC FRONT SECTION.

O ADD SUBSTANCE TO SIDE OF AR, PAINT ROCKER PANEL (SECTION BELOW DOOR LINE) IN BODY COLOR.

UNCTIONAL-LOOKING WHEELS ENHANCE HILOSOPHY AND LOOKS OF CAR.

CONVERTIBLE TOP HAS TIGHT AND PERFECT FIT AND IS AESTHETICALLY PLEASING.

OR THIS SIZE CAR, SMALLER ND LESS BULKY SIDEVIEW MIRRORS WOULD LOOK BETTER AND MORE INTEGRATED.

BACK-UP LIGHTS APPEAR TOO BIG IN RELATION TO SIZE OF CAR.

HARD-EDGED LICENSE PLATE RECESS DOES NOT MATCH CAR'S SOFT LINES. PROPOSE GRADUAL CAVITY AS SHOWN.

IT'S A WINNER!

MAZDA RX-7 GTU^{S}

The Rotary Club welcomes a new member

YOU'VE SEEN THE concept from other manufacturers: Chevrolet offers the Camaro RS, looking to the uninitiated like an IROC-Z but powered by a standard-equipment 2.8-liter V-6 engine. It's the General's answer to buyers looking for a little panache without the penalties of exorbitant insurance and fuel-consumption rates. Porsche has given us the "Turbo-look" option, where those smitten with wide fender flares and whale-tail spoilers can have them fitted to a normally aspirated 911. Presto! A bargain Turbo-imitator with enough money left over to buy a whole case of those trick titanium Porsche Design cigarette lighters. Now Mazda has seen fit to offer another version of its venerable RX-7–the GTU^{S}–endowed with all the sartorial splendor and handling prowess of the top-of-the-line Turbo, without any sort of forced breathing.

From a styling standpoint, Mazda has left the svelte good looks of the

PHOTOS BY SCOTT DAHLQUIST

RX-7 intact (after all, we'll see a completely new car in late 1991 or early 1992) and seems content to fool with minor styling changes that freshen up the car's looks and provide Mazda salesmen with an answer to the inevitable question, "How is it different from the 1988 model?" Viewed from the front, there are subtle contour changes to the flexible bumper cap, and the rectangular cooling air intake loses its horizontal slat. At the rear, the RX-7 line receives smoked taillight lenses covering dual round lights, much like those seen on last year's Ford Thunderbird. The alloy wheels, redesigned for lighter weight, are another element that smacks of FoMoCo design: If you squint just a little bit, they could be the optional wheels right off a Mercury Sable.

Under the GTU[S]' weight-saving aluminum hood (also found on Turbo and Convertible models) nests the powerplant synonymous with the RX-7 designation: a 1308-cc twin-rotor Wankel. For 1989, this little beer keg of an engine develops more power (160 bhp, 14 more than last year's) through an improved intake system, higher compression ratio (9.7:1, up from 9.4:1), lighter rotors (by 5 percent), a quicker-reacting engine-management computer and a less restrictive airflow meter. And those lighter rotors, when coupled to a 16-percent-lighter flywheel, allow the engine's redline to be raised by 1000 rpm to a dizzying 8000.

While peak torque is increased by only 2 lb.-ft. (to 140), the torque range is slightly broader, thanks to variable intake geometry utilizing a rotary valve (no pun intended) connecting the intake tracts. The valve opens at about 4500 rpm, effectively shortening the length of the tracts to take advantage of the different pressure-wave pulse effect present at high rpm. Just think of it as a poor man's supercharger.

Once rolling, acceleration feels brisk, helped by a low 4.30:1 final drive ratio. But 140 lb.-ft. of torque doesn't move nearly one and a half tons of car away from rest with the

■ Enormous tach takes center stage—significant in light of the twin-rotor's rev capability.

■ The interior strikes a nice balance between sportiness and luxury. Seats' side bolstering copes well with car's lateral acceleration capability; their support frees one to steer with the wheel instead of using it as a grab handle. Twin-rotor 1308-cc Wankel gets more power and torque for 1989.

MAZDA RX-7 GTUs

0–60 mph .... 8.6 sec
0–¼ mi .... 16.7 sec
Top speed .... est 130 mph
Skidpad 0.81g
Slalom .... 63.8 mph
Brake rating .... very good

PRICE

List price, all POE	**$19,600**	Price as tested	**$20,708**

Price as tested includes std equip. (AM/FM stereo/cassette, elect. window lifts), air cond ($859), cruise control ($249).

ENGINE

Type	**2-rotor Wankel**
Displacement	79.8 cu in./1308 cc
Chamber width x major axis	3.15 x 9.45 in./ 80.0 x 240.0 mm
Compression ratio	9.7:1
Horsepower (SAE):	**160 bhp @ 7000 rpm**
Torque	**140 lb-ft @ 4000 rpm**
Maximum engine speed	8000 rpm
Fuel injection	Mazda elect. port
Fuel	unleaded, 87 pump oct

GENERAL DATA

Curb weight	**2775 lb**
Test weight	**2925 lb**
Weight dist, f/r, %	**50/50**
Wheelbase	95.7 in.
Track, f/r	57.1 in./56.7 in.
Length	169.9 in.
Width	66.5 in.
Height	49.8 in.
Trunk space	11.3 + 7.8 cu ft

DRIVETRAIN

Transmission .... **5-sp manual**

Gear	Ratio	Overall ratio	(Rpm) Mph
1st	3.48:1	14.94:1	36
2nd	2.00:1	8.61:1	62
3rd	1.37:1	5.87:1	91
4th	1.00:1	4.30:1	125
5th	0.76:1	3.25:1	est (6310) 130

Final drive ratio 4.30:1
Engine rpm @ 60 mph in 5th 2910

CHASSIS & BODY

Layout	**front engine/rear drive**
Body/frame	aluminum hood, unit steel
Brakes, f/r	**10.9-in. vented discs/ 10.8-in. discs,** vacuum asst
Wheels	cast alloy, **16 x 7J**
Tires	Bridgestone Potenza RE71, **205/55VR-16**
Steering	**rack & pinion,** power asst
Turns, lock to lock	2.7
Suspension, f/r:	**MacPherson struts,** A-arms, coil springs, tube shocks, anti-roll bar/**multilink** (trailing arms with camber-control links, semi-trailing links, floating hubs), coil springs, tube shocks, anti-roll bar

ACCELERATION

Time to speed	Seconds
0–30 mph	2.9
0–60 mph	8.6
0–80 mph	15.3
Time to distance	
0–100 ft	3.4
0–500 ft	9.0
0–1320 ft (¼ mi)	16.7 @ 84.5 mph

BRAKING

Minimum stopping distance	
From 60 mph	152 ft
From 80 mph	256 ft
Control	very good
Pedal effort for 0.5g stop	22 lb
Fade, effort after six 0.5g stops from 60 mph	22 lb
Brake feel	excellent
Overall brake rating	very good

FUEL ECONOMY

Normal driving	19.0 mpg
EPA city/highway	17/25 mpg
Fuel capacity	18.5 gal.

INTERIOR NOISE

Idle in neutral	48 dBA
Constant 70 mph	73 dBA

HANDLING

Lateral accel (200-ft skidpad)	0.81g
Balance	moderate understeer
Speed thru 700-ft slalom	63.8 mph
Balance	moderate understeer

Subjective ratings consist of excellent, very good, good, average, poor.

authority of, say, the 2.7-liter Porsche 944. Launched with a touch of wheelspin, the GTUs will see 20 mph in 1.6 seconds, 40 in 4.4, 60 in 8.6, 80 in 15.3.

Although it won't be embarrassed in the daily traffic nip-and-tuck, the GTUs is happiest on an open stretch of road with its tachometer needle spinning in the upper third of its rev range. At 7000 rpm, the rotary engine doesn't even seem to be breaking a sweat, because each rotor is turning at only one-third the speed of the flywheel. And the smoothness is uncanny; the engine's mechanical pitch and exhaust-note ferocity increase as the revs climb, but the low level of vibration stays nearly constant.

Once you've built up that head of steam and find the corner you thought was constant radius suddenly starting to tighten, you'll be glad the GTUs has many of the Turbo's handling and braking mechanical pieces. Although the Turbo's ABS isn't available, premium hardware such as aluminum 4-piston brake calipers and 10.9-in. vented discs up front (also found in the GXL and Convertible models) and 205/55VR-16 tires (lesser models make do with 60-series tires of the same width but of 15-in. diameter) help to haul this RX-7 down in short order. Lay into the brake pedal and you'll be rewarded with a reassuringly firm, progressive feel. In our simulated panic stops, the system was easily modulated and wasn't prone to lockup.

For an "Ortega Flog"—office jargon for a pleasure drive along Southern California's Ortega Highway, a sinuous mountain road that links San Juan Capistrano with the city of Lake Elsinore—the RX-7 GTUs would be one of our top choices. Thick anti-roll bars (0.95 in. front,

Test Notes . . .

■ **A better revver than bruiser, the GTUs can require a tricky blend of revs and clutch timing off the line, but stopping is a total pleasure, the pedal being superbly sensitive and communicative.**

■ **Steered smoothly into a bend, the RX-7 moderately understeers, but if you saw at the steering wheel, the tail will dance out—never truly threatening, but sometimes unexpectedly.**

0.55 in. rear) keep body roll to a minimum and are made more direct and effective this year by the substitution of ball joints for rubber bushings on their end links. The exceptionally rigid unit body and the power-assisted rack-and-pinion steering serve as great conduits of information for what's happening at the tire's contact patch. In Chassis-Speak, the GTU^{S} is a smooth, fast talker, and by listening to what it has to say, you'll find it easy to probe its high limits with confidence.

Minor revisions to the interior make a very good driving environment even better. Round ancillary gauges, to the left of the huge centered tachometer, replace the previous pie-wedge ones. The already supportive seats receive additional bolstering, and a new pistol-grip shift lever nicely complements the thick-rimmed steering wheel; both are swaddled in high-grade leather.

For those who like to manipulate throttle and brake simultaneously with one foot (the infamous heel-and-toe technique), the relationship between the pedals is just right.

Big rubber knobs on either side of the instrument pod operate the lights and the windshield wipers; the knobs' action is crisp but it's difficult to see the mode symbols, displayed through tiny windows. The ventilation system is effective and easy to use, with logically placed vents, straightforward labeling and levers that slide with well-oiled precision. Only the very tall staff members lodged a major complaint about the interior: marginal head room. If you're much over 6 ft. tall, expect either to bump your head every once in a while or to put in an order for the Convertible.

The RX-7 continues to be a strong seller in the 2-seater sports-car market, a tribute to a car whose design is now four years old. Perhaps it's because the RX-7 is more elemental and direct than Toyota's Supra, has a cleaner, less fussy design than Mitsubishi's Starion, and is substantially less expensive than anything Porsche has to offer. Maybe the rotary engine's uniqueness appeals to people who march to a different drummer. In any case, the GTU^{S} is a welcome addition to Mazda's lineup. If the Miata's not your cup of tea, the base GTU is simply too base and you're repeatedly backhanded by the Good Hands People when trying to insure the Turbo, the GTU^{S} may just fill the bill.

Decisions, Decisions...

Convertible has standard 15-in. BBS wheels; sure-fire way to tell the Turbo from the GTU^{S} is the former's prominent hood scoop.

THERE IS A sumptuous variety about the New England weather that compels the stranger's admiration – and regret, Mark Twain once told the members of the New England Society, at a dinner speech 113 years ago. Had Mr. Twain been a sports-car enthusiast (and were alive today), he would have noted a similar ambivalence toward the 1989 RX-7 lineup.

There's a sumptuous variety of models–the base GTU (replacing last year's SE for least expensive honors), the GTU^{S}, the luxury-oriented GXL, the Convertible and the Turbo–and in their own ways, they're all admirable. The regrettable part is that the enthusiast of average means will have to settle for just one of them!

Mazda sent us three of the five models for testing and driving impressions: the GTU^{S} (see accompanying test), a Turbo with a 5-speed manual and a Convertible with a 4-speed automatic.

For the wind-in-the-face crowd, there's the Convertible, with a base price of $25,600. Surprisingly, through some aerodynamic fine-tuning, Mazda has kept cockpit wind-buffeting to a minimum. By simply lifting the Windblocker (a small hinged panel behind the seats), turbulence is reduced to the point that it's easy to hold a conversation without shouting; you emerge from a long drive without feeling pummeled by the roundhouse punch of swirling air.

The electrically actuated top is simple and easy to use, and it incorporates a heated glass rear window. Of course, the top can be lowered completely, but, if desired, you can remove the rigid targa panel and drive the car in what Mazda calls the half-top mode, with the portion aft of the doors remaining upright.

Thanks to a stiffened chassis structure, the dreaded cowl shake is noticeable only on very bumpy roads. Thicker steel stampings (used for the rocker panels, A-pillars and drivetrain tunnel) in conjunction with cross-members installed behind the engine, seats and between the rear spring towers give the Convertible about 70 percent of the hardtop's torsional rigidity.

The downside of this extra beef (some 250 lb. more than a comparably equipped GTU) is somewhat lethargic acceleration: When fitted with the 4-speed automatic transmission, the drop-top RX-7 takes 11.3 seconds to reach 60 mph; the quarter mile comes up in a leisurely 18.4 sec., with a terminal speed of 78.5 mph.

If the Convertible's sluggishness casts a pall over your otherwise sunny outlook, the $25,950 Turbo's scintillating performance should throw you into fits of laughter. A Twin-Scroll turbocharger, new for 1989, gives the exhaust charge from each rotor a clear shot at the turbo's impeller, improving response and fattening the torque curve across the board, especially at low rpm.

With assistance from a new electronically controlled wastegate, lighter rotors and a bump in compression to 9.0:1 (up from 8.5:1), the engine develops 200 bhp and 196 lb.-ft. of torque, and shot-puts this RX-7 to 60 mph in a scant 6.3 sec., with the turbo making a sound like a New Year's Eve noisemaker gone mad. The quarter-mile marker is a blur as it flashes past in 15.1 sec., accomplished at 93.5 mph.

These are numbers that best those generated by Toyota's supercharged MR2 and compare favorably to cars like the Ford Mustang GT and the Ferrari 328 GTS. Meaty 205/55VR-16 tires and firm spring rates aid in putting all that energy to the ground, and standard ABS helps undramatically and safely to rein it all back in.

It's a tough choice. If we could just get Mazda to make a Turbo Convertible model . . .—*Douglas Kott*

MIATA MANIA

From overnight sensation to Rat Racer Roadster

BY RICHARD HOMAN

PHOTOS BY JEFFREY R. ZWART

THE MYTHOLOGY SURROUNDING the Mazda MX-5 Miata is already being created. Frightful stories of motorcycle cops pulling Miata drivers over just to talk to them and check out their cars. Stories of owners refusing even to take their Miatas out in public because they're weary of answering questions about the car from other drivers on the road and from excited crowds that gather wherever the little convertible is parked.

Much of the Miata folklore revolves around people who want to purchase a Miata and can't. We have heard stories about impatient West Coast Mazda customers, facing sold-out showrooms and waiting lists that extend for months, who

traveled to Colorado in hopes of taking immediate delivery of their Miatas from Denver dealers (at $3500 above the list price). And then there is the apocryphal tale of the man who walked into a Santa Monica, California Mazda dealership with $25,000 cash in his pocket, offered it for the dealer's demo Miata and was turned down.

The groundswell of anticipation for the Miata, spawned by the motoring press last spring, erupted in July and August into a news-media blitz. Television, radio and general-interest publications were spending unprecedented amounts of air time and ink on "the hot, new sports car from Japan." Interest in the Miata expanded well beyond the usual enthusiast channels and jumped all the demographic fences. Miata Mania had set in.

It was during the convertible-perfect months of late summer and early fall of this year that the first batch of these 2-seat, front-engine/rear-drive lightweight Japanese roadsters finally found its way into the anxious hands of owners who had deposited their hearts and money. Suddenly, the heaviest burden of all—that of fulfilling inflated public expectations—had been placed on this lovely lamb of a car by the media's lionizing.

Well, the Miata delivered, in spades, when it came down to the public reckoning. As an exception to Robert Louis Stevenson's rule that "traveling hopefully is better than arriving," the arrival of the Miata was better than anyone dared hope for. The car had pert performance and road manners, plus looks and a personality second to no other car on the highway. The high school letterman, the NASA engineer, the cosmetologist and the retired Air Force colonel each got the satisfaction he or she expected from top-down touring in a Miata.

The only signs of a backlash appeared during the early months of the Miata's availability and were directed toward dealers, all of whom were deluged with orders for Miatas. And some of whom were selling the $14,000 to $16,000 car for as much as $10,000 more than that.

Miata Mania isn't showing many signs of waning, however. One telephone-bound Mazda Information Bureau representative reported receiving an average of 20 Miata inquiries a day. And Norman Garrett, founder of the Miata Club of America, estimates that by the end of the year his organization will have well over 5000 members. And it won't be a fraternity: Garrett says that the Miata Club's roster is 20-percent female, outstanding for a car club.

Clearly, then, the Mazda MX-5 Miata is this year's rage, turning heads anywhere the sun shines upon an open road. The challenge now, of course, is to come up with something that

will make a Miata *owner* do a double take. Enter Rod Millen Motorsport and the Turbo MX-5 Miata.

"Personally, I was thrilled when I heard that we were going to put on a turbocharger and do the suspension kit. Eventually, I wish to own this car myself. So doing this turbocharger gives me the opportunity to get a little advanced knowledge about it."

Rolf Samulewicz is the "chief fabricator/part-time mechanic" at Rod Millen Motorsport. Samulewicz is one of the ablest soldiers in an able outfit. In past lives, he did fabricating and wrenching for a number of Grand National Sports Truck teams in the Mickey Thompson stadium racing series, ending up as Rod Millen's crew chief in the series. When the team disbanded, Samulewicz stayed with Millen, working in the New Zealander's Huntington Beach, California shop, helping prep Rod's hugely successful Mazda rally cars and swinging the aftermarket hammer on the anvil for Rod Millen Motorsport.

Having only joined Millen Motorsport at the beginning of this year, Samulewicz is also the junior member of Millen's 10-man crew, but his involvement and dedication (and his acceptance of all-nighters as standard operating procedure in motorsports) typify the team's enthusiasm, experience and confidence.

■ **Aluminum heat shield hides turbocharger and bundle-of-snakes header welded up from many curved sections of stainless-steel tubing. Shock-tower brace shores up chassis for the much higher loads it encounters. Quality of installation and fabrication is first-rate.**

"When [Millen team manager and modifications designer] Murray Thomas came up to me and said, 'Okay, we're going to do a turbo on the Miata,' I just . . . like a horse . . . I was raring to go and I've been going ever since."

When Rod Millen, Murray Thomas and company looked at the aftermarket potential of a stock Miata, they decided, not surprisingly, that the cadre of Miata owners who drive with a little extra bravura needed cars with more power and tighter handling, plus some cosmetic body pieces for distinction. They envisioned a street-racer, to be built as an image car for Rod Millen Motorsport and Mazda, and wanted to identify their company "as the leader in aftermarket performance kits for Miatas."

In preparing to increase the performance potential of the Miata's engine, the Millen Motorsport crew had a significant head start. Already closely involved with Mazda because of the racing programs, Millen and his men were able to watch as the Miata developed, speculating about what they could do for the car. And on the engine side of things, the team was able to draw heavily from its experience

■ **Unidirectional 15 x 7-in. wheels, manufactured by Panasport exclusively for Millen Motorsport, are shod with meaty Bridgestone RE71Rs. The combination fills out the wheel wells perfectly and contributes largely to the car's 0.90g skidpad performance.**

with a similar 1.6-liter dohc 16-valve fuel-injected powerplant of Mazda origin—that of the turbocharged 323 GTX, a longtime bread-and-butter Millen rally car.

"We retained as much of the original Miata engine as we felt we could, and we've combined that with the 323 GTX engine. Our intention was to do this kit and have it look as close to a factory installation as possible. I think we've gone a long way to achieve that."

That's Richard Kelsey speaking. Kelsey is the head of sales at Rod Millen Motorsport and the organizer of the Turbo MX-5 Miata project. He is also a Mazda rally driver of some excellence, having broken the record for fastest time at the Pikes Peak Hill Climb in the GT Production class last year in a 323 GTX (Kelsey's Pikes Peak record was broken this year by none other than Rod Millen driving a Mazda MX-6).

On the way to making its Miata, Millen Motorsport raided the 323 GTX's engine, pirating the car's camshafts, pistons (thus lowering the Miata's compression ratio from 9.4:1 to 7.2:1), and radiator, turbocharger and intercooler.

The turbo implant is the golden nugget of the Millen Miata's performance upgrade. Tucked tightly into the engine bay, the little turbo is mounted stage left up against a Millen Motorsport-designed, stainless-steel exhaust manifold. The intercooler is fitted neatly into the nose of the Miata, just ahead of the radiator. This design allows air to be ducted from the turbo to the intercooler in a clean, straight path. Air routing from the intercooler into the intake assembly is a bit more protracted, but still efficient. After finishing its turbine-spinning duties, the exhaust is let out into a 2.5-in. stainless-steel exhaust system and glass-pack muffler.

The original intention in all this was to set the turbo's boost at 16 psi and bump the engine's output from 116 bhp to just a little over 200. But rather than bumping, the Millen modifications drop-kicked the Miata's horsepower up to 230, two nags shy of doubling it. Not bad for a mere 35-lb. weight increase.

With the turbo shrouded under a highly effective (and highly necessary) heat shield, it all makes for a fairly elegant installation, with absolutely minimal repositioning of stock pieces within the engine bay.

"We wanted to maintain the motif of the motor compartment that Mazda created. We did that by snugging up the turbo in a position that would allow the stock components to remain where they were," Samulewicz explained. "And that's all part of the heat shielding: to protect the air box and the other vital components in the motor. Plus the turbo's well hidden, so those people who don't like to have their car's potency exposed won't. I come from a big street-racing background, and you don't want to have all your secret parts out in the open."

In a tip of the hat to Mazda, the Millen Motorsport designers chose to stay with the stock Miata gearbox. With all that extra power on tap, however, they felt it necessary to replace the stock clutch with a special unit designed by Centerforce. At the other end of the drivetrain, they installed one of their own rally-competi-

PRICE

	Millen Miata	Stock Miata
Base price	**$25,612***	**$13,800**

GENERAL DATA

	Millen Miata	Stock Miata
Curb weight	**2240 lb**	**2205 lb**
Test weight	**2390 lb**	**2365 lb**
Weight dist, f/r, %	**51/49**	**52/48**
Wheelbase	89.2 in.	89.2 in.
Track, f/r	56.4 in./56.9 in.	55.5 in./56.2 in.
Length	155.4 in.	155.4 in.
Width	65.9 in.	65.9 in.
Height	46.3 in.	48.2 in.

ENGINE

	Millen Miata	Stock Miata
Type	turbo 4-valve/cyl dohc **inline-4**	4-valve/cyl dohc **inline-4**
Displacement	97 cu in./1597 cc	97 cu in./1597 cc
Bore x stroke	3.07 x 3.29 in./ 78.0 x 83.6 mm	3.07 x 3.29 in./ 78.0 x 83.6 mm
Compression ratio	7.2:1	9.4:1
Horsepower	**est 230 bhp @ 5500 rpm (dyno)**	**116 bhp @ 6500 rpm (SAE)**
Torque	**est 235 lb-ft @ 5000 rpm**	**100 lb-ft @ 5500 rpm**
Maximum engine speed	7000 rpm	7000 rpm
Fuel delivery	Nippondenso elect. port	Nippondenso elect. port
Fuel	prem unleaded, 92 pump oct	unleaded, 89 pump oct

DRIVETRAIN

	Millen Miata	Stock Miata
Transmission	**5-sp manual**	**5-sp manual**

Gear	Millen Ratio	Millen Overall ratio	Millen (Rpm) Mph	Stock Ratio	Stock Overall ratio	Stock (Rpm) Mph
1st	3.14:1	13.48:1	33	3.14:1	13.48:1	33
2nd	1.89:1	8.12:1	54	1.89:1	8.12:1	54
3rd	1.33:1	5.72:1	77	1.33:1	5.72:1	77
4th	1.00:1	4.30:1	102	1.00:1	4.30:1	102
5th	0.81:1	3.50:1	est (7000) 126	0.81:1	3.50:1	est (6510) 117

	Millen Miata	Stock Miata
Final drive ratio	4.30:1	4.30:1
Engine rpm @ 60 mph in top gear	3340	3340

CHASSIS & BODY

	Millen Miata	Stock Miata
Layout	**front engine/rear drive**	**front engine/rear drive**
Body/frame	unit steel & fiberglass	unit steel
Brakes, f/r	**10.9-in. vented discs/ 10.8-in. vented discs,** vacuum asst	**9.3-in. vented discs/ 9.1-in. discs,** vacuum asst
Wheels	cast alloy, **15 x 7**	cast alloy, **14 x 5½JJ**
Tires	Bridgestone Potenza RE71R **205/50VR-15**	Bridgestone SF-325 **P185/60HR-14**
Steering	**rack & pinion**	**rack & pinion**
Turns, lock to lock	2.8	2.8
Turning circle	na	30.0 ft
Suspension, f/r	**upper & lower A-arms,** coil springs, tube shocks, anti-roll bar/**upper & lower A-arms,** coil springs, tube shocks, anti-roll bar	**upper & lower A-arms,** coil springs, tube shocks, anti-roll bar/**upper & lower A-arms,** coil springs, tube shocks, anti-roll bar

*Base price for the **Millen Miata** includes the following modifications: oversized brakes ($2240), intercooler and plumbing ($2200), fuel-management system ($1100), limited-slip differential ($750), 15 x 7-in. alloy wheels ($740), turbo ($650), 205/50VR-15 tires (est $650), stainless steel exhaust manifold ($550), heavy duty clutch ($425), exhaust temp gauge ($350), larger anti-roll bars ($350), stainless steel exhaust system ($330), camshafts ($300), strut tower brace ($225), Momo steering wheel ($210), spring set ($200), rear spoiler ($195), rear deck aero cover ($195), pistons ($120), boost gauge ($32).

■ **Stiffer springs and thick Heim-jointed anti-roll bars keep the chassis flat around corners; huge rotors and calipers (from an RX-7 Turbo donor) haul it down from speed in confidence-inspiring fashion.**

tion limited-slip differentials.

According to Richard Kelsey, the Miata was only too willing to cooperate with the engine/drivetrain modifications. The usual niggling snags that always seem to occur were noticeably absent on this effort. The biggest hassle the project faced was adding a second engine-management computer to handle fuel-delivery concerns—flows and mixtures; an administrative problem at worst, Kelsey said.

That big power demands big brakes is a point not lost on the folks at Rod Millen Motorsport. They have adapted the brake assembly from Mazda's RX-7 Turbo for use on their Miata. That means 10.9-in. vented discs with 4-piston calipers at the front wheels and 10.8-in. vented discs at the rears. For added efficiency, Millen Motorsport drills the rotors and uses semi-metallic competition brake pads.

Aluminum mounting bars had to be fashioned to mount the RX-7 discs on the Miata hubs, and special brackets were made to adapt the RX-7 calipers for mounting as well.

When it came time to address the issue of handling, the suspension upgrades selected were simple, yet extremely effective. Millen's geometry squad installed beefier anti-roll bars—1.0 in. front, 0.75 in. rear—with urethane bushings. Stiffer springs went in and the car was lowered 1.0 in. in the nose and 0.75 in. in the tail. For good measure, a transverse reinforcing brace was mounted across the engine and bolted to the shock towers. Impressed

ACCELERATION		
Time to speed	Seconds	Seconds
0–30 mph	2.0	3.0
0–60 mph	6.4	9.5
0–80 mph	10.8	16.5
Time to distance		
0–100 ft	3.0	3.4
0–500 ft	8.0	9.2
0–1320 ft (¼ mi)	14.7 @ 95.5 mph	17.0 @ 81.5 mph
BRAKING		
Minimum stopping distance		
From 60 mph	131 ft	154 ft
From 80 mph	235 ft	270 ft
Control	very good	very good
Pedal effort for 0.5g stop	19 lb	16 lb
Fade, effort after six 0.5g stops from 60 mph	19 lb	22 lb
Brake feel	excellent	excellent
Overall brake rating	excellent	good
HANDLING		
Lateral accel (200-ft skidpad)	0.90g	0.83g
Balance	mild understeer	moderate understeer
Speed thru 700-ft slalom	63.6 mph	62.4 mph
Balance	mild oversteer	mild oversteer
FUEL ECONOMY		
Normal driving	est 15.0 mpg	25.5 mpg
Fuel capacity	11.9 gal.	11.9 gal.
INTERIOR NOISE		
Idle in neutral	57 dBA	52 dBA
Constant 70 mph	76 dBA	77 dBA

Subjective ratings consist of excellent, very good, good, average, poor; na means information is not available.

again with the quality of the canvas they were painting on, the Millen Motorsport clan stayed with the Miata's stock shocks.

Slung lower and filled out by the special 15 x 7-in. Rod Millen Motorsport directional aero wheels with 205/50VR-15 Bridgestone Potenza RE71R tires, the Miata was beginning to look tougher; it was getting an attitude.

So, being swept up in Miata Mania as much as anybody, we thought, "Let's ask them to paint the car yellow and we'll take it for a drive!" And that we did. The car fired up and idled with a smooth, clear and low Mills Brothers rumble. The throttle response wasn't quite as quick as we were used to on the stock Miata. The unassisted steering, with the extra tire width to direct, was slightly heavy at low speeds.

Then we tried the throttle. After only the slightest delay, the turbo was up and singing, the power band stretched like it had no ceiling. Up through the gears, the pop-off valve spewed its exhalation between shifts. Normally a 9.5-second-to-60-mph car, the Millen Miata reached the mark in 6.4 sec. When the quarter-mile sign blipped past (14.7 sec.), we were at 95.5 mph and climbing. That's Ferrari 328, Porsche 944 S2 and Mustang GT territory.

PHOTO BY RICHARD M. BARON

■ **Fat-section Momo steering wheel is all the better to tell you what the front tires are doing; boost gauge and exhaust-temperature pyrometer keep driver apprised of engine's goings-on. Spoiler and wider wheels and tires give the illusion of flared fenders, but it's just plain ol' Miata sheet metal.**

Turbo lag? Yes, there was some, but no more than with several production cars we can name. And once you've made your way past 3000 rpm in 1st gear and you're on boost, it's very easy to keep the Turbo Miata in the power range through the rest of the sprint.

The ride is a bit stiff, making everyday use possible, but not preferable. The Miata's 0.90g skidpad number, on the other hand, is unquestionably "stand-up-and-take-a-bow" time. And the big RX-7-derived brakes make for short, excellent stopping figures.

The question of how to customize the Miata was, in many ways, the most delicate issue of all facing the Millen crew. How do you add your signature to an almost universally adored design?

It doesn't matter whether the decision was made to err in the direction of conservativism, or good taste simply prevailed. With all the screeching add-ons and tarty accessories that fertile, unrestrained human minds in the aftermarket have come up with, it is to Millen Motorsport's credit that the Miata not only survived, but prospered in the customizing ritual.

The sum of the Millen Miata's body parts is a slight rear spoiler, an attractive rear-deck cover—both beautifully integrated into the design—and a mesh grille at the mouth of the car.

To sit in the car is to sit in a stock Miata with only a few exceptions. At the center in front of the shifter, a boost gauge and a pyrometer, measuring exhaust temperature, suggest something other than a normal Miata. The comfortable, supportive seats are retained, but the steering wheel is a sculptured Momo (before they could install the Momo wheel, the Millen Motorsport bomb squad drew lots to see who was going to defuse the airbag on the stock steering wheel). Millen also tried out a Momo shift knob, but the superb, tapered knob designed for the Miata's short-throw gearbox had it beat.

And hence the open road beckons the driver of the Turbo MX-5 Miata, a magnificent sports car, bred in the spirit and bred in the flesh.

But certainly there is a price to be paid for this much exhilaration. The engine and drivetrain modifications go for around $7000; the brake system will run you about $2300; the suspension goodies, $775; the wheels are $185 apiece; the spoiler and deck cover, $195 each. And these are install-it-yourself prices.

Too much? Well, friend, these are frenzied times, fraught with Miata Mania. And, as Mazda's new icon has taught us, enthusiasts will pay big money for fun and exclusivity if they get plenty of both. So our philosophy is: Any time you can drop 3 full seconds off your 0 to 60 time *and* flabbergast a Mazda Miata owner—what are you waiting for?

Lesser Evils

WERE MILLEN Motorsport's Turbo Miata an airplane, it would be a jet fighter, I'd say the General Dynamics F-16. They are both small, agile, blindingly fast and require a pilot not afraid to "push the envelope." And like most projects done under military contract, the Miata's price—approaching $26,000—may seem unfathomable to the enthusiast whose wallet is as thin as the atmosphere at 50,000 feet.

While it's possible to buy one, this muscular little tarmac-dart is intended primarily as a showcase and test bed to parade the firm's capabilities. For performance-seekers of lesser means, there's encouraging news—both Millen's firm and Mazdasports (a new division formed by Oscar Jackson Racing of Honda- and Acura-modifying fame) are developing full lines of engine and suspension parts from which the shade-tree mechanic can assemble a flatter-cornering, more responsive Miata.

On rather short notice, both firms put together Miatas featuring just suspension pieces, wider 15-in. wheels and fatter tires, and stiffer, linear-rate coil springs that lower ride height by approximately one inch. Stock anti-roll bars were tossed out in favor of larger-diameter bars that ride in urethane bushings and have adjustable end links. While the stock Miata is one of the finer devices for hustling down a twisty road, the modified versions make the production car feel slightly wobble-kneed by comparison.

We sampled the Mazdasports Miata first. The well-scuffed Yokohama A008Rs adhered to the 200-ft.-diameter skidpad with a vengeance to generate 0.89g, 7 percent more grip than the stock Miata on the same surface. At the cornering limit, its balance was commendable; a drop-throttle stunt that would have had the driver of the stock car cranking in lots of countersteer required only a minor steering correction in the Mazdasports Miata to keep everything tidy and under control.

The red Millen Motorsport car was the lateral-g champ. With its wider wheels and shaved Bridgestone Potenza RE71Rs, it clung tenaciously to its line around the skidpad to a seat-bolster-compressing 0.94g. Around the track, it didn't quite have the nimbleness and agility of the Mazdasports Miata, but the distinction is a very subtle one. Its adhesion-limit behavior, too, was exemplary; a bit more neutral than the Mazdasports car, it would undramatically slide all four tires in unison when the gas pedal was lifted.

Both companies are in the early stages of their engine-development programs and already offer headers and exhaust systems. (Contact Mazdasports at 16291 Gothard St., Huntington Beach, Calif. 92647; Rod Millen Motorsport, 7552 Reynolds Circle, Huntington Beach, Calif. 92647.) Experimentation continues with various camshaft-lobe profiles in an effort to eke out more performance from Mazda's willing twin-cam 1.6—a few more beans will be the perfect complement to their chassis' new-found athleticism.

—Douglas Kott

SPECIFICATIONS

	Stock Miata	Millen Motorsport Miata	Mazdasports Miata
Front anti-roll bar diameter	0.75 in.	1.00 in.	0.94 in.
Rear anti-roll bar diameter	0.47 in.	0.75 in.	0.63 in.
Wheels	14 x 5½	15 x 7	15 x 6½
Tires	Bridgestone SF-325 P185/60HR-14	Bridgestone Potenza RE71R 205/50VR-15	Yokohama A008R 195/50VR-15
Lateral acceleration	0.83g	0.94g	0.89g

Rod Millen Motorsport Miata.

Stock Miata.

Mazdasports Miata.

MAZDA

RX-7 TURBO

Twinkle, twinkle, little Wankel, now we know how well you crankel

PHOTOS BY SCOTT DAHLQUIST

PEERING INTO THE engine bay, you can't get a good look at it, but you know it must be there. It's nestled way back near the firewall for good weight distribution, topped by a sizable intercooler and hides behind the alternator, power steering pump and a maze of tubing and ducting.

It's the thing responsible for one of the sweetest automotive success stories of all time, and it's what makes the RX-7 unique. The reclusive guest of honor is the 2-rotor Wankel, a tiny oak barrel of an engine that has aspects of both turbine and 2-cycle engines wrapped into one tidy aluminum and cast-iron package.

It has no valvetrain; the intake charge enters when the rotor uncovers a port, and the exhaust leaves in the same manner. And like a 2-stroke, minute amounts of oil are mixed with the fuel to keep engine internals well lubed. (In the Wankel's case, the oil is to help the rotor's seals live long, healthy lives.) Among its other attributes are light weight (compared to engines of similar power) and its un-

■ With the leather interior option, the RX-7 Turbo could easily masquerade as one of Europe's finest. And with Mazda's floating-hub rear suspension, it has the handling grace to match its appearance.

canny smoothness, due in part to the rotors' relatively slow rotation in relation to crankshaft speed.

Thanks to the unfaltering vision of Mazda Chairman Kenichi Yamamoto and his following of talented engineers, the rotary is still with us in the RX-7. It faced near-eradication in the early Seventies because of its tendency to slurp exorbitant amounts of fuel. When the Arabs started tightening the petroleum spigot, Americans looked toward more efficient piston-engine cars, and sales of the Wankel-powered RX-3 dropped away like a Stuka dive bomber en route to target.

Lucky for us enthusiasts, Mazda has been beavering away ever since, making substantial improvements in fuel economy, driveability, reliability and power. A turbocharged version of the 1308-cc Wankel was first offered in 1986, producing the very respectable figure of 182 bhp. For 1990, the engineers have outdone themselves, bumping the pressurized engine's output to 200 bhp at 6500 rpm and torque to the tune of 196 lb-ft at 3500, 13 more lb-ft than last year's.

The increase was accomplished with a little fine tuning. Lighter rotors (trimmed by 5 percent) and flywheel (16 percent lighter) reduce the rotating mass for quicker response. Shallower recesses in the rotors raise the compression ratio from 8.5:1 to 9.0:1. Through a bit of microchip wizardry, the computing speed of the engine-management computer has been doubled, and the turbo engine now breathes easier through a less-restrictive throttle body and a new twin-scroll turbocharger.

This last variation capitalizes on the extremely "pulsey" nature of the rotary's exhaust flow. Instead of combining the exhaust pulses into a single opening before they hit the impeller's blades, the plumbing is such that each port's full-strength exhaust pulse gets a clear shot at the impeller which, according to Mazda, results in reduced turbo lag and increased torque.

According to our collective throttle foot and the test-track acceleration numbers, Mazda is right on the money. Boost is felt as a powerful push in the lower back just shortly after the tach needle swings past the number 2 on its ascent to the 7; and 60 mph comes up in a scant 6.4 seconds, just a wink after the 2-3 shift. It takes the RX-7 Turbo a paltry 15.0 seconds to trip the quarter-mile timing lights, enough sustained squirt to eclipse such performance benchmarks as the Taurus SHO, the Firebird Trans Am GTA and the supercharged MR2.

When driving smoothness is at a premium—say, when your mother-in-law is balancing a cup of coffee in her lap—the turbo seems a bit over-responsive.

That's fine if Kenny Bernstein or Joe Amato is the principal pilot, but there's a glitch that surfaces when the nature of the drive doesn't involve getting the lowest E.T. When driving smoothness is at a premium—say, when your mother-in-law is balancing a cup of coffee in her lap—the turbo seems a bit over-responsive. To maintain steady acceleration, you've got to use some finesse and back out of the throttle a bit. That may seem to be a little picky, but other manufacturers (Mitsubishi pops to mind) have produced less-fussy turbocharged engines of similar output.

The RX-7's chassis, however, needs no apologies; only a little constructive criticism. It uses MacPherson struts (employing beautiful forged-aluminum lower A-arms) up front and a modified trailing-arm setup in back. The rear suspension is reactive, its toe-in settings increasing in relation to side-load and braking forces.

There's a tradeoff with this arrange-

ment, though. At high cornering loads, it sacrifices a little of its communication to the driver for what chassis engineers deem the greater good: goof-proof stability to keep Bondurant-dropout Ned Neophyte clear of the guardrail. Despite this minor character flaw, the chassis has a sure-footed, tight feel that results from lots of grip from its Goodyear Gatorbacks (size 205/55VR-16, on 16 x 7-in. cast-alloy wheels) and minimal body roll, thanks to fat anti-roll bars (this year with ball-jointed end links for better precision). Brakes are exemplary, with large vented discs, front 4-piston aluminum calipers and ABS cooperating to yield some of the shortest stopping distances we've recorded.

The RX-7's interior also lives up to the promise of the engine and chassis, with supportive, big-bolstered seats that don't complicate ingress and a full complement of gauges both nice to look at and easy to read quickly. Like the unit-steel body they're housed in, all the major controls have a solid, rock-of-Gibraltar feel—the firm brake pedal, the robust, pistol-grip gearshift lever and the thick, leather-covered steering wheel whose column doesn't flex even when you're flailing away at it in a tight corner. It's as if Mazda hired the Samsonite gorilla to conduct interior durability tests. The ventilation controls are first-cabin, with slide levers that move with well-oiled precision and mode buttons that click with easy authority.

No one will ever laud the RX-7's exterior design as being original—there's a lot of similarity to a certain sports car from *das Vaterland*—but it's aging well. From its steeply sloped nose (with its little flash-to-pass lenses and prominent hood scoop nicely breaking up the monotony) to its smoked rear taillight lenses covering dual round lights (new for this year), the RX-7 continues to have a strong visual presence. There are fresher sports-car designs in the marketplace, like the Nissan 300ZX, that are sure to attract more covetous glances. There are cars that offer nearly the same performance for substantially less money; the Ford Mustang GT and Mitsubishi Eclipse/Eagle Talon pop to mind. But what they can't match about the RX-7 Turbo is something that isn't immediately apparent; you have to open up the Mazda's hood and poke around for awhile.

—Douglas Kott

PRICE

List price, all POE .... $26,530   Price as tested $27,530

Price as tested includes std equip. (air cond, AM/FM stereo/cassette, cruise control, central locking, elect. window lifts, elect. adj mirrors, rear window heat, ABS, leather seats ($1000)

ENGINE

Type	turbo 2-rotor Wankel
Displacement	1308 cc
Bore width x chamber major axis:	80.0 x 240.0 mm
Compression ratio	9.0:1
Horsepower, (SAE):	200 bhp @ 6500 rpm
Torque	196 lb-ft @ 3500 rpm
Maximum engine speed	7000 rpm
Fuel injection	electronic
Fuel	unleaded, 87 pump oct

GENERAL DATA

Curb weight	est 2985 lb
Test weight	est 3135 lb
Weight dist, f/r, %	50/50
Wheelbase	95.7 in.
Track, f/r	57.1 in./56.7 in.
Length	169.9 in.
Width	66.5 in.
Height	49.8 in.
Trunk space	6.5 cu ft

DRIVETRAIN

Transmission .... 5-sp manual

Gear	Ratio	Overall ratio	(Rpm) Mph
1st	3.47:1	14.23:1	29
2nd	2.00:1	8.20:1	57
3rd	1.37:1	5.47:1	82
4th	1.00:1	4.10:1	114
5th	0.72:1	2.95:1	est (6300) 140

Final drive ratio 4.10:1

Engine rpm @ 60 mph in 5th 2640

CHASSIS & BODY

Layout	front engine/rear drive
Body/frame	unit steel
Brakes, f/r	10.9-in. vented discs/ 10.8-in. vented discs, vacuum assist, ABS
Wheels	cast alloy, 16 x 7J
Tires	Goodyear Eagle VR, 205/55VR-16
Steering	rack & pinion, power assist
Turns, lock to lock	2.7
Suspension, f/r:	MacPherson struts, lower A-arms, coil springs over tube shocks, anti-roll bar/trailing arms with camber control arms on drop links, semi-trailing links, floating hubs, coil springs, tube shocks, anti-roll bar

ACCELERATION

Time to speed	Seconds
0–30 mph	2.2
0–40 mph	3.5
0–50 mph	4.8
0–60 mph	6.4
0–70 mph	8.7
0–80 mph	10.9
0–90 mph	14.1
0–100 mph	17.3
Time to distance	
0–100 ft	3.1
0–500 ft	8.1
0–1320 ft (¼ mi)	15.0 @ 93.1 mph

BRAKING

Minimum stopping distance	
From 60 mph	124 ft
From 80 mph	225 ft
Control	excellent
Overall brake rating	excellent

FUEL ECONOMY

Normal driving	est 16.0 mpg
EPA city/highway	16/24 mpg
Fuel capacity	18.5 gal.

HANDLING

Lateral accel (200-ft skidpad)	0.85
Speed thru 700-ft slalom	na

Subjective ratings consist of excellent, very good, good, average, poor. na means information is not available.

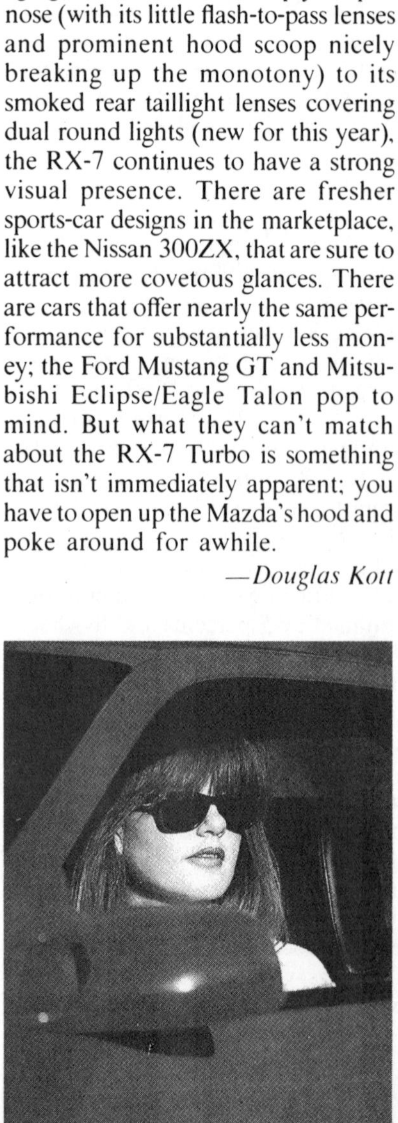

MX-5 MIATA

Top-down, wind-in-the-hair motoring in light, agile cars as pleasing to the eye as they are to drive. This is the very essence of sports car motoring, resurrected by Mazda. It's a new car, but the feeling is familiar, like discovering a long lost friend.

Summertime, and the shiftin' is easy

PHOTOS BY LESLIE L. BIRD & DAVID W. BIRD II

CAN THE MAZDA MX-5 Miata turn water into wine? No, but it *has*, by and large, turned automotive journalists into jelly. Can it part the Red Sea? No, but it can pave a road across the Pacific Ocean to deliver motoring enthusiasts out of the desert and into the Promised Land of Sports Cars.

Can it update the dream of the inexpensive, lightweight, 2-seat convertible? Yes. Can it predict the future? Let's hope so.

More than any other car in recent memory—and even fairly distant memory—the Miata has created a stir of almost embarrassing proportions.

What could turn an otherwise jaded group of PC-wielding car writers into a bunch of gushing quill pushers, all romance and flowers? What is it about this car that makes drivers jump out of their cars at stoplights to inquire, "What is it? Who makes it? How much?" Or, further, to have an in-the-know driver follow you for blocks just to tell you, "I've got one on order!"

The answer, of course, is fun. Pure, simple, undistilled, clear-eyed driving fun; from the top of your head (a convertible lid) to the tips of your toes (a drilled accelerator pedal).

It was during the middle of this decade that some of Mazda's young Turks decided that what this world needed

was a fun, bargain-priced, honest, classic sports car. Today, the idea seems as obvious as an elephant in your living room, especially when the 1989 denizens of the sports-car fun zone are gone (Pontiac Fiero, Bertone X1/9), expensive (the next-generation Toyota MR2 and Mazda's own RX-7) or front-drive (Honda CRX Si). And the only classic roadster of the bunch, the Alfa Romeo Spider, was 10 years old at least 10 years ago.

Mazda thought it was getting a

■ If the engine were mounted sideways in the engine bay, one could mistake this car for the 323 GTX.

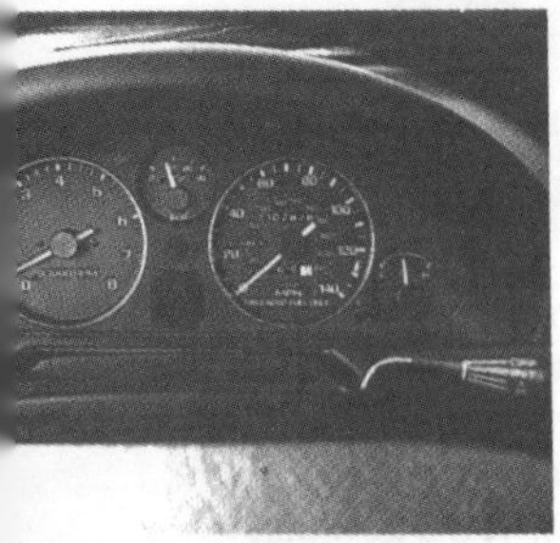

niche car: a $14,000, front-engine/rear-drive, open 2-seater. What the Japanese automaker has tapped into, however, is the recognition of a longing in the U.S. for inexpensive, lovable roadsters; a thirst that has gone unslaked for a decade. Factor in the reliability of Japanese manufacturing, and it is easy to see why the Miata is being treated like an avatar instead of an automobile.

But we tested it as an automobile. And as such, we discovered that the Miata is a great car with good, but not great, performance. We also found that while numbers don't lie, the whole of a car is often greater than the sum of its sums.

At 2205 lb, the Miata is about 100 lb heavier than a CRX Si and 150 lighter than the 1989 MR2—two good cars for comparison, because they, like the Miata, are powered by 1.6-liter, 4-valve-per-cylinder engines. No matter, both the Toyota and the Honda are significantly quicker off the line (0–60 mph) and back onto the dime (braking from 60 and 80 mph) than the Mazda.

The Miata's fuel-injected 116-bhp dohc four (the only engine Mazda offers) keeps giving all the way up to its 7000-rpm redline. Musically, the engine is a perfect tenor, serenading the driver lustily to 6200 rpm, where the exhaust note begins to sour.

This engine is mated exclusively (no automatic transmission offered) to a terrific 5-speed manual with super-short throws and a precise feel.

Fairly short gearing in 1st and 2nd gears gives the driver the sensation of being able to launch the car quickly when the light turns green. The P185/60HR-14 Bridgestones also contribute to this feeling of quickness as they allow some off-the-line wheelspin in 1st and a tiny chirp in 2nd.

But none of this changes the stock MX-5 Miata's best 0–60-mph sprint, which, at 9.5 seconds, is a least a full second slower than that of either the MR2 or the CRX Si.

The Miata's 4-wheel disc brakes don't seem to give it any relative stopping advantage, either. The brake feel is very good, however, with excellent communication between the pedal, brakes and tires. And while we would have hoped for braking distances of something shorter than the 270 ft it took the lightweight Miata to stop from 80 mph, we were impressed by the sensitivity of the system.

The suspension setup on the MX-5 Miata is simple and direct, front and rear: upper and lower A-arms, coil springs over tube shocks, and an anti-roll bar. Mazda has set up the stock Miata for comfort more than handling. The ride is comfortably soft with more body roll than we would have expected.

All of this is not to say that the Miata, with its light weight and short wheelbase, isn't quick and responsive; it's both. But it could be quicker and even more responsive. Already it is easy to see the aftermarket explosion this car is going to cause (for an example, see the accompanying article). A tighter suspension and wider tires here, a few engine modifications there, and you could have the premier autocrosser of the Nineties.

So much for the stuff of engineers, now let's go to the poet's camp. The Miata's exterior looks as if it were grown from nature, not carved in a wind tunnel. Brazenly, the Miata flaunts its high coefficient of drag.

That natural theme is carried through to the inside of the Miata as well. There are very few sharp edges to cut the eyesight on and no complication to clutter the mind (and precious little trunk space to accumulate any clutter *in*). The controls are laid out simply, as are the chrome-bezeled gauges on the instrument panel.

The driver also fits into this environment as a natural consequence. The seats are comfortable for distance driving and supportive during the twisting fun runs. While just a tad shorter on leg room than an MR2, the Miata will still fit a 6-ft 4-in. editor as if he were born to the car.

We even had the convertible top up one time to check on the head room (we found plenty). The one morning when we did get a bit of rain, it was a simple matter to manually raise the top, zip up the plastic rear window and clamp the two latches into their seats in the windshield header. The top's sealing was superb.

With the top down, the swirling winds wafted freely through the driving compartment, but did so without buffeting the occupants to distraction or noisily ruining conversation. It is also important to point out that chassis flexing, a crucial consideration on convertibles, was virtually nonexistent in the Miata. Driver ego flexing, on the other hand, was rampant.

It is comforting to note that as a result of Mazda's determination to keep it simple and inexpensive, the MX-5 Miata is an easy car to keep track of in your mind. As noted earlier, there is only one engine/gearbox combination. Three colors are available—red, white and blue—with a hot yellow (as seen on the Club Racer version at the Chicago auto show) and British Racing Green (if you really want to rub it in) being considered for the future.

If you must have your sports-car cake and eat it too, two optional upgrade packages are available: Package A includes alloy wheels, an AM/FM stereo/cassette system, leather-wrapped steering wheel and power-assisted steering. Package B includes all the add-ons of package A plus the sins of cruise control, headrest speakers and—gasp—electric window lifts. Additionally, there are four straight-ahead options: a limited-slip differential, a compact disc player, air conditioning and a detachable hardtop (available in red only now).

This road test of the Mazda MX-5 Miata is not a test of numbers. The objective performance results are here. Some sunny day, however, you'll put on your driving gloves, take the top down and head out to your favorite country road. You'll run through the gears, upshifting and downshifting, as one with the road, feeling out your favorite corner as though it were your first time.

At that moment, all of those numbers will just get blown away in the rushing wind like so many useless gasoline receipts in the passenger seat. And the smile on your face will be real. And the twinkle in your eye as you wave a gloved hand to your brethren, other Miata drivers, will be earnest and enthusiastic.

—Road & Track, November 1989

PRICE

List price, all POE .... $13,800 Price as tested.................... est $15,740

Price as tested includes: option package A (power-assist steering, AM/FM cassette, alloy wheels, leather steering wheel), $1145, air cond ($795)

ENGINE

Type	dohc 4-valve inline-4
Displacement	1597 cc
Bore x stroke	78.0 x 83.6 mm
Compression ratio	9.4:1
Horsepower, SAE net:	116 bhp @ 6500 rpm
Torque	100 lb-ft @ 5500 rpm
Maximum engine speed	7000 rpm
Fuel injection	Nippondenso electronic port
Fuel	unleaded, 91 pump oct

GENERAL DATA

Curb weight	2205 lb
Test weight	2365 lb
Weight dist, f/r, %	52/48
Wheelbase	89.2 in.
Track, f/r	55.5 in./56.2 in.
Length	155.4 in.
Width	65.9 in.
Height	48.2 in.
Trunk space	6.0 cu ft

DRIVETRAIN

Transmission .... 5-sp manual

Gear	Ratio	Overall ratio	(Rpm) Mph
1st	3.14:1	13.48:1	33
2nd	1.89:1	8.12:1	54
3rd	1.33:1	5.72:1	77
4th	1.00:1	4.30:1	102
5th	0.81:1	3.50:1	est (6510) 117

Final drive ratio 4.30:1

Engine rpm @ 60 mph in 5th 3340

CHASSIS & BODY

Layout	front engine/rear drive
Body/frame	unit steel
Brakes, f/r	9.3-in. vented discs/ 9.1-in. discs, vacuum assist
Wheels	cast alloy, 14 x 5½JJ
Tires	Bridgestone SF-325, 185/60HR-14
Steering	rack & pinion, power assist
Turns, lock to lock	2.8
Suspension, f/r:	upper & lower A-arms, coil springs, tube shocks, anti-roll bar/upper & lower A-arms, coil springs, tube shocks, anti-roll bar

ACCELERATION

Time to speed	Seconds
0–30 mph	3.0
0–40 mph	4.8
0–50 mph	7.0
0–60 mph	9.5
0–70 mph	12.3
0–80 mph	16.5
0–90 mph	21.6
Time to distance	
0–100 ft	3.4
0–500 ft	9.2
0–1320 ft (¼ mi)	17.0 @ 81.5 mph

BRAKING

Minimum stopping distance	
From 60 mph	154 ft
From 80 mph	270 ft
Control	very good
Overall brake rating	good

FUEL ECONOMY

Normal driving	25.5 mpg
EPA city/highway	est 25/30 mpg
Fuel capacity	11.9 gal.

HANDLING

Lateral accel (200-ft skidpad)	0.83g
Balance	moderate understeer
Speed thru 700-ft slalom	62.4 mph
Balance	mild oversteer

Subjective ratings consist of excellent, very good, good, average, poor.

COMPARISON
R&T
ROAD TEST

PHOTOS BY JOHN LAMM

MERCURY CAPRI versus MAZDA MX-5 MIATA

Sporty car meets sports car

AUSTRALIA: LAND OF kangaroos, koalas and Ken Done clothes. Shrimps on barbies and *Crocodile Dundee.* And now, Capri.

You know, as in the automobile.

Not the sporty sedan built by Ford of Europe in the Seventies, or the rebodied Mustang built by Mercury in the Eighties, but a new 2+2 convertible that's fun to drive yet practical, fully equipped yet affordable. Styled by Italians, built by Australians using mechanicals made in Japan.

That's Capri III, a small roadster comparable to the Mazda Miata in concept (fun driving in an open-top tourer) and in price. But not necessarily in size, performance, handling, comfort, room and that elusive something that separates a sports car from one that is merely sporty.

Sports car: That's the Miata to a T. Hot off the assembly line, the little car stole our hearts as it scampered off with one of *Road & Track's* "World's Best" honors, in the fast company of rich playmates such as the Ferrari Testarossa, Corvette ZR-1 and Porsche 911 Carrera 4. "A refreshing throwback to simpler times," we called it, describing a car that has become the standard against which other enthusiast automobiles are measured. The new Capri is one of these contenders, and while Lincoln-Mercury spokesmen point out that their roadster is not meant to go head-to-head with the Miata, it's difficult not to make this comparison.

That Miata and Capri are different is obvious. The MX-5 is a traditional, front-engine/rear-wheel-drive design while the Mercury is a front-driver. Also, the Capri looks bigger. But is it? Yes—and no. At 166.1 in. overall, the Mercury, with its 94.7-in. wheelbase (5.5 in. greater than the Mazda's), is nearly 11 in. longer than the Miata. And 2.0 in. taller, top up. However, because of its lean, Italian-styled flanks, the XR2 is actually 1.3 in. narrower than the well-rounded, California-styled MX-5, whose track is 0.6 in. wider up front and 0.2 in. wider at the rear.

No question about which car is heavier. The 2610-lb. XR2, the extensively optioned, turbocharged version of the Capri, outweighs our test Miata by 405 lb. But then, the Mercury carries a lot more gear than the Mazda—for example, turbocharger with intercooler, driving lights, rear spoiler, electric window lifts and door locks, electrically powered side mirrors, cruise control and, of course, a rear seat (of sorts).

Actually, the differences between the two cars are as much philosophical as they are physical. Mazda, starting with the proverbial clean sheet of paper, chose to produce a contemporary version of the classic Fifties' sports car. Ford, drawing upon existing concepts, sought to develop a fun runabout with wide-ranging appeal (and numerous creature comforts), using existing componentry.

The idea had been germinating for nearly a decade. In the early Eighties, FoMoCo had considered building an affordable 2-seat convertible based on the Barchetta, Ghia's 1983 Frankfurt show car—using European Ford Fiesta underpinnings. But the Euro-roadster fared poorly in customer clinics (Americans didn't like

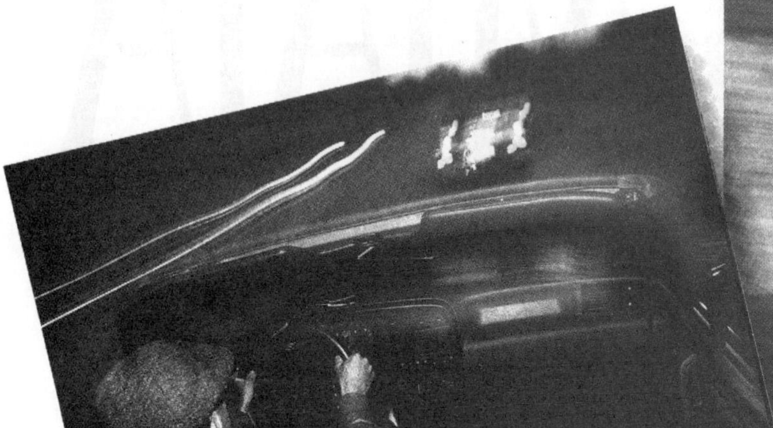

Whether it's the Capri or the Miata, putting the top down can only raise the excitement level. Stroker Ace cap is optional.

its radical lines and its diminutive size). And the Europeans decided the car would be too expensive to produce.

At about this time, Ford's Asia-Pacific Automotive Operations (APAO) was looking for ways to expand its export market and stepped in. APAO hired Italdesign to build a concept car based on the Australian Laser platform, displayed it and an enlarged Barchetta at a clinic in Los Angeles (alongside the Pontiac Fiero, Honda CRX, Nissan EXA and Bertone X1/9) in 1984.

Consumers—expressing a desire for convenience, practicality and affordability in a car that was fun to drive—sent Ford back to its stylists, namely Italdesign and Ghia, who developed two new concept cars for another Ford clinic held in San Diego in mid-1985.

This time, both cars caught the public's fancy—the Ghia for its exterior, the Italdesign for its interior—and APAO incorporated the best of each into the production Capri. Asia-Pacific had already struck a deal with Mazda to supply engines, mechanicals and floorpans to the Australians. Although Mazda's own sports car, the Miata, was in the works, executives of both companies saw no conflict in the two projects.

With Ford of North America as its primary customer (for 30,000 out of 40,000 cars annually), the Capri program began in earnest.

PHOTOS BY JOE RUSZ

BARCHETTA: The Little Boat That Never Saled

Before Miata, before Capri, but after Austin-Healey, Triumph and MG, there was the Barchetta. In Italian, it means "little boat," an apt name for a jaunty runabout that sought to bring the same joy of open-air cruising to landlubbers.

To Ford, "Barchetta" meant an inexpensive sports car that could be built using an existing platform, engine/drivetrain and mechanicals. Back in 1982 when the folks at Ford were first inspired by this "better idea," the chassis of choice was the Ford Fiesta—because of its size and front-wheel drive.

From Dearborn the word went out. To Ghia where designers were told to develop a car that would have modern appeal, but would not be elaborate; that would be solid, fun, roadworthy and affordable.

After chopping the front cowl and shortening the wheelbase 2–3 in., Ghia craftsmen cloaked the Fiesta chassis with a crisply styled steel body. They installed two seats, a manual top—and set the car aside.

A German automotive magazine published photos of the Barchetta and received more than 1000 letters urging Ford to build it. Ford wanted further reassurance, so the roadster was displayed at the 1983 Frankfurt salon where it received wide acclaim. But unbeknownst to all of us, the winds of change had blown the little boat off course.

Other design studies, leading to the Capri, swamped the Barchetta. But some of its lines, especially its distinctive Ghia nose and twin-bubble hood, survive and serve as both a reminder of things past and a possible precursor of things to come.

—Joe Rusz

■ Capri's engine (above) is the same turbocharged, intercooled 1.6-liter inline-4 used in Mazda's 323 GTX, something of an off-the-shelf rally car. The Miata's 1.6-liter naturally aspirated four makes 116 bhp, 16 fewer than the Capri's, but makes up in eagerness for what it lacks in power.

Giorgetto Giugiaro's Italdesign craftsmen were brought in once again, this time to make running changes and to build prototypes that were completed in early 1987 and shipped to the U.S., Europe and Australia for testing and further refinement. The first Capri pilot rolled off Ford's Broadmeadows assembly line in June 1988, 15 months before the initial batch of production Capris, which were earmarked for Australia.

Unlike the Miata, with its purpose-built chassis, the Capri is based on the pre-1990 Mazda 323 platform that until this year was shared with Ford of Australia's Laser sedan. No worries, mate, because the MacPherson-strut front/Chapman-strut rear design shared by all three cars has been tuned to suit the sporting nature of the Capri—basically by stiffening the springs, shocks and anti-roll bars. And because the car is a convertible, numerous stiffening gussets, braces and reinforced body panels (double-wall doorsills, for example) have been incorporated into the unit body to lend rigidity and to minimize cowl shake.

Like the floorpan and other running gear, the Capri powerplant is also Mazda-sourced. It's the same turbocharged 1598-cc dohc inline-4 used by the 323 GTX—coupled to the same 5-speed manual transmission with identical gear and final drive ratios. Using 9.0 psi of boost, the little twincam develops 132 bhp in turbocharged form (100 bhp in normally aspirated, base model trim). That puts the XR2 16 horses up on the Miata with expected results: The turbo Capri is 1.2 seconds quicker from 0 to 60 mph and 0.7 sec. quicker in the quarter mile.

Disc brakes with 10.2-in.-diameter vented rotors up front, 8.7-in. solid rotors at the rear, round out the list of mechanicals supplied by Mazda. When pressed—hard—as in stops from 80 and 60 mph to zero, you'll find that the

XR2's binders perform very well, even if the Mercury's stopping distances (affected significantly by its not-too-grippy tires) are a mite longer than the Miata's.

Compared with the Miata's cozy cockpit, the Capri's Giugiaro-styled cabin is capacious. The widely adjustable seats are supportive, yet comfortable, especially on long trips, and there's a marginal rear seat that makes the Capri a 2+2—or serves as a nice adjunct to the XR2's roomy trunk, which has a trapdoor that can be opened to allow long objects such as skis to protrude into the passenger compartment with the seatback folded down. If there's a discordant note, it's Ford's choice of materials and fabrics, which give the interior a look that is not at the top of its class.

The Mercury's dash, analog instruments and controls are straightforward, as are the Miata's. Headlight, driving light and air-conditioning

PHOTOS BY WM A. MOTTA

Fare Dinkum

"Keep left except to pass." No worries, mate. Life (driving) in the fast lane is like that for those of us who travel with the hammer down. Except that in this case, left means right, because in Australia, which is south of the equator (where water flows counterclockwise down the drain), they drive ON THE WRONG SIDE OF THE ROAD.

Not a problem in a right-hand-drive automobile. Except DCI West Coast Magazines Design Director William A. Motta and I were making this two-day trek from Albury to Sydney in a U.S.-spec Mercury Capri XR2 with the steering wheel on the opposite side of the car.

A glance down the left shoulder and, gulp, she'll be right.

Okay, so there was that time I made a turn onto a secondary road and instinctively veered right. But that Ocker in the Nissan Patrol with the roo bars saw us and even gave us a proper Aussie greeting, complete with flashing headlights, blaring horn and jaunty salute.

Some short stints on Australian Interstates served as the only reminders that about 7 million people (out of 16 million) inhabit the countryside traversed during our 527-mile jaunt. Mostly, we motored along in solitude: through the rolling, California-like hills of Victoria (The Garden State) to Canberra; into the Pennsylvania-like mountains and along the Hawaii-like coast of New South Wales (The First State) to Sydney (new official slogan, "You're beautiful").

■ A removable hardtop, one of the Capri's few options, adds some extra rigidity and sells for $1274.

Midway, there was "brekky" with Senator John Button in Australia's $2 billion Parliament House (their Capitol Building). The event brought out the Aussie news media—even though I informed them that if nominated, I would not accept, if elected, I would not serve.

Early on, there were refreshments—at Alf and Maija Wilson's quaint "hotel" in Tintaldra (population 18 with one on the way). Maija introduced us to "squash" (actually lemon squash), a carbonated drink somewhere between Squirt and Mountain Dew. Goes by the trade name Solo if you want to order by mail.

Later in the day, lunch exposed us to "snags" (sausages) grilled on a barbie, along with steaks ("steaks"), salads, pâté and homemade desserts. And served alfresco by David and Marianne Hutchison—alongside a trout-filled lake in the Snowy Mountains. Ah, wilderness!

In the course of our journey, we hung our hats in Hyatts—in Melbourne (their Chicago) and Canberra (Washington, D.C.)—and rested in the Regent, in Sydney by the bay (San Francisco). We watched footie (Australian-rules football) on the telly (Sony) and drank Foster's from tinnies (cans). And got tucker-ed out (stuffed full of food).

Australia: It's not all that different from America, mate. Except when it comes to driving, in which case you're not wrong if you're not on the right.

—Joe Rusz

switches are located on the dash, while wiper controls are stalk-mounted, next to a grippy, fat steering wheel.

Both cars are equipped with manually operated folding tops. The Miata's, which is smaller, can be raised practically from one's seat; the Capri's is handled best while standing alongside the car. Pop open the rear cowl, pull up on the top, but don't latch it until you've secured the backlight, which hooks to a small molding. Then, fasten the top to the windshield header. The XR2's top is quite rigid and this contributes to the car's quietness.

When it comes to ergonomics, you'll find that both Mazda and Ford have done their homework. In short, all primary controls, pedals, etc. are properly placed and, as we're wont to say, "the gearshift falls readily to hand." However, the Miata's has a precise feel that's missing in the Capri's.

Preciseness: That's the difference between a sports car and a sporty car. And to a large extent, the difference between Miata and Capri.

Drive them back-to-back, and you'll find that the Capri is faster in a straight line and a kick to drive when the turbo is spooled up. The XR2's 16-valve four is a spirited engine that gets to full boost quickly and with nary a whoosh. But rowing up or down through the gears isn't always as much fun as it is in the Miata. That's because the MX-5's gear ratios are better matched and the shifter is more direct, while the XR2 shifter (though solid and requiring a firm hand) feels slower, a bit vague and doesn't like to be rushed.

Then there's the matter of torque steer. Although Ford chassis/suspension engineers have attempted to minimize it by going so far as to mount the engine/gearbox subframe directly to the unit body without using rubber mounts, some of us found that under certain conditions (particularly on boost), the XR2's steering was adversely affected. And yet, one man's torque steer is another man's torque tightening. Or as one driver said, "The car doesn't change direction, just feel."

Throw in some twisty bits, and you'll find the Miata in its element. Steering feel and input are good and turn-in is crisp. Partly because of its shorter wheelbase, but mostly because of its nearly perfect 52/48 weight distribution, the Miata tracks faithfully. It exhibits moderate understeer in slow corners and gets increasingly neutral in fast ones. Sticky tires help, especially on the skidpad where the MX-5 pulls an impressive 0.83g.

Throw in the same twisties, and you'll find the Capri capable, but anxious (moderately soft suspension makes it feel wallowy). Like many fwd automobiles with much of their weight up front, the XR2 understeers, albeit mildly, in slow and fast turns. Its original-equipment tires aren't particularly grippy—witness its mediocre 0.78g showing on the skidpad—although they suffice in the slalom where front-wheel drive saves the day—and enables the Mercury to reach 62.0 mph, a tick slower than the Mazda.

Cruise the freeways or make that daily commute, and you'll find that sporty car beats sports car. The Capri rides more smoothly, not only because its more supple suspension absorbs bumps more readily, but also because the XR2's longer wheelbase reduces porpoising. The Mercury is quieter because its more powerful engine doesn't work as hard and because the turbocharger muffles exhaust noise. And because the top fits really well (an optional hardtop makes the car even quieter).

Now, the moment of decision. Which car belongs in your driveway?

For those who will settle for nothing less than a pure sports car (and are willing to live with the consequences), only the Miata will do. The Mazda's attractive styling, intelligent design, impeccable road manners (but not necessarily its straight-line performance and room) are what make the MX-5 a best-seller.

For those who want to enjoy the same pleasure of open-air motoring (and are willing to live with different consequences), it must be the Capri. The Mercury's styling, while bland to some folks, is generally appealing—with top down. The car is tidy and trim, and has lots of interior room. And a host of creature comforts. Ride is on the soft side, but the XR2 can be pushed reasonably hard without embarrassing itself or the driver. Although it's not a sports car like the Miata, it's certainly a welcome addition to the sporty, top-down crowd.

PRICE

	Mazda MX-5 Miata	Mercury Capri XR2
List price, all POE	**$13,800**	**$15,522**
Price as tested	**$15,740**	**$15,522**
Price as tested includes	option package A (includes AM/FM stereo/cassette, alloy wheels, leather steering wheel) $1145, air cond ($795).	Price as tested includes std equip. (air cond, AM/FM stereo/cassette, elect. window lifts, elect. adj mirrors, central locking, cruise control).

GENERAL DATA

	Mazda MX-5 Miata	Mercury Capri XR2
Curb weight	**2205 lb**	**2610 lb**
Test weight	**2365 lb**	**2760 lb**
Weight dist, f/r, %	**52/48**	**63/37**
Wheelbase	89.2 in.	94.7 in.
Track, f/r	55.5 in./56.2 in.	54.9 in./56.0 in.
Length	155.4 in.	166.1 in.
Width	65.9 in.	64.6 in.
Height	48.2 in.	50.2 in.

ENGINE

	Mazda MX-5 Miata	Mercury Capri XR2
Type	4-valve/cyl dohc **inline-4**	turbo 4-valve/cyl dohc **inline-4**
Displacement	97.4 cu in./1598 cc	97.4 cu in./1598 cc
Bore x stroke	3.07 x 3.29 in./ 78.0 x 83.6 mm	3.07 x 3.29 in./ 78.0 x 83.6 mm
Compression ratio	9.4:1	7.9:1
Horsepower (SAE)	**116 bhp @ 6500 rpm**	**132 bhp @ 6000 rpm**
Torque	**100 lb-ft @ 5500 rpm**	**136 lb-ft @ 3000 rpm**
Maximum engine speed	7000 rpm	6500 rpm
Fuel injection	Nippondenso elect. port	Nippondenso elect. port
Fuel	unleaded, 87 pump oct	unleaded, 87 pump oct

CHASSIS & BODY

	Mazda MX-5 Miata	Mercury Capri XR2
Layout	**front engine/rear drive**	**front engine/front drive**
Body/frame	unit steel	unit steel
Brakes, f/r	**9.3-in. vented discs/9.1-in. discs,** vacuum asst	**10.2-in. vented discs/8.7-in. discs,** vacuum asst
Wheels	cast alloy, **14 x 5½JJ**	cast alloy, **14 x 5½JJ**
Tires	Bridgestone SF-325, **P185/60HR-14**	Dunlop SP Sport D8, **P185/60R-14 82H**
Steering	**rack & pinion,** power asst	**rack & pinion,** power asst
Turns, lock to lock	2.8	3.2
Turning circle	30.0 ft	32.5 ft
Suspension, f/r	**upper & lower A-arms,** coil springs, tube shocks, anti-roll bar/ **upper & lower A-arms,** coil springs, tube shocks, anti-roll bar	**MacPherson struts,** lower A-arms, coil springs, tube shocks, anti-roll bar/ **Chapman struts,** dual lateral links, trailing links, coil springs, tube shocks, anti-roll bar

DRIVETRAIN

Transmission: Mazda MX-5 Miata **5-sp manual**; Mercury Capri XR2 **5-sp manual**

Gear	Miata Ratio	Miata Overall ratio	Miata (Rpm) Mph	Capri Ratio	Capri Overall ratio	Capri (Rpm) Mph
1st	3.14:1	13.48:1	33	3.31:1	12.73:1	33
2nd	1.89:1	8.12:1	54	1.83:1	7.06:1	59
3rd	1.33:1	5.72:1	77	1.23:1	4.75:1	88
4th	1.00:1	4.31:1	102	0.97:1	3.73:1	112
5th	0.81:1	3.50:1	est (6510) 117	0.80:1	3.06:1	est (5900) 124

	Mazda MX-5 Miata	Mercury Capri XR2
Final drive ratio	4.30:1	3.85:1
Engine rpm @ 60 mph in 5th	3340	2855

ACCOMMODATIONS

	Mazda MX-5 Miata	Mercury Capri XR2
Seating capacity	2	2+2
Head room, f/r	37.5 in.	38.0 in./32.0 in.
Seat width, f/r	2 x 19.0 in.	2 x 20.5 in./44.0 in.
Front leg room/rear knee room	43.0 in.	41.0 in./18.0 in.
Trunk space	6.0 cu ft	7.5 + 7.2 cu ft

FUEL ECONOMY

	Mazda MX-5 Miata	Mercury Capri XR2
Normal driving	25.5 mpg	25.0 mpg
EPA city/highway	25/30 mpg	23/28 mpg
Fuel capacity	11.9 gal.	11.1 gal.

INTERIOR NOISE

	Mazda MX-5 Miata	Mercury Capri XR2
Idle in neutral	52 dBA	49 dBA
Constant 70 mph	77 dBA	76 dBA

HANDLING

	Mazda MX-5 Miata	Mercury Capri XR2
Lateral accel (200-ft skidpad)	0.83g	0.78g
Balance	moderate understeer	mild understeer
Speed thru 700-ft slalom	62.4 mph	62.0 mph
Balance	mild oversteer	mild understeer

ACCELERATION

	Mazda MX-5 Miata	Mercury Capri XR2
Time to speed	Seconds	Seconds
0–30 mph	3.0	2.7
0–60 mph	9.5	8.3
0–80 mph	16.5	14.4
Time to distance		
0–100 ft	3.4	3.4
0–500 ft	9.2	8.8
0–1320 ft (¼ mi)	17.0 @ 81.5 mph	16.3 @ 85.0 mph

BRAKING

	Mazda MX-5 Miata	Mercury Capri XR2
Minimum stopping distance		
From 60 mph	154 ft	162 ft
From 80 mph	270 ft	278 ft
Control	very good	very good
Pedal effort for 0.5g stop	16 lb	25 lb
Fade, effort after six 0.5g stops from 60 mph	22 lb	25 lb
Brake feel	excellent	very good
Overall brake rating	good	good

Subjective ratings consist of excellent, very good, good, average, poor.

CARRYING

"HOW COME YOU guys are all driving the same kind of car?" a woman asked us at a gas station on the north coast of California.

We turned and looked at our little mobile sports-car collection, which we'd been flogging through the hills and down the rocky coastline for three days. The same kind of car?

Okay, four of them were red, granted, and she had probably not noticed the black 300ZX was in our group. And from the side the RX-7 did have a wedgy resemblance (not entirely unintended) to the Porsche 944. Also, the Miata, though Elan-like from the front, had a side profile not shockingly different from the other two. The MR2 was not even a convertible, however, and had a certain son-of-Testarossa flair to at least its front two-thirds. The rear, however, dissolved into a bobtailed shape similar to the others. And it, too, was bright red. In addition, there was not much difference in the size of the cars—inches separated the shortest (Miata) from the longest (RX-7).

So I suppose a civilian on the street could be forgiven for thinking all of us belonged to some weird club for an unnamed marque of red sports car, but we still had a good grin over the idea that

THE TORCH

A look at the naturally aspirated, base-level (sometimes affordable) sports cars that define the genre for the Nineties: Mazda Miata, Mazda RX-7, Nissan 300ZX, Porsche 944, Toyota MR2

BY PETER EGAN

PHOTOS BY RICHARD M. BARON & DEAN SIRACUSA

they were all the same model. After our hundreds of miles of driving and swapping cars, each had taken on a marked personality. We were driving the cars, in fact, to see what traits separated them and to find out what your money gets you these days in naturally aspirated, standard-model sports cars.

Our group had gathered at the San Francisco airport, some of us flying in to pick up cars from distributors and others driving up from Southern California. We spent a night in San Francisco, during which a couple of the younger editors were introduced to the traditional North Beach delights of browsing at the City Lights bookstore after downing too much espresso at Cafe Trieste, the combination of which allows you to read good books all night long. In the overcast morning we crossed the Golden Gate Bridge into Marin County and blitzed north along the brooding, beautifully empty coast toward Bodega Bay. During the next three days we drove, and then drove some more, over winding coastal range roads that are essentially too good to be true, writing notes all the while.

Here are a few impressions and conclusions from our travels, beginning alphabetically with the Mazdas:

Mazda MX-5 Miata

ONE EDITOR HAD the nerve to claim that this was the only real sports car in the group. Why? Because it is not carrying any unnecessary baggage, he explained. No vestigial rear seats, not an inch of unneeded width or length, and at 2205 lb. it's the lightest car in the group. It has no more bulk, weight or horsepower than it needs to get the job done.

And what is that job? For most of us, the job of a sports car is to take pavement—particularly winding pavement—and alchemize it into that elusive thing called fun. This the Miata does better than all the others. Its quick steering, willing engine, short-throw gearbox, taut suspension and fits-like-a-glove driving position give it a feeling of instantaneous response that conjures up memories of the old Lotus Elan, only this time the headlights work and the water pump lasts quite a bit longer.

There is some small payment to be made for the Miata's spirited agility, however. It's a little stiffer and higher-strung than the others, and probably least suitable for a long cross-country trip if comfort and quietness are what you require. The snarly exhaust and less compromised suspension make for a greater sense of fatigue at the end of an eight-hour driving day. That doesn't mean that any of us would hesitate to drive it from Alaska to Florida, however.

Some drivers also mentioned that the steering is a little too quick right around center. This trait, which makes the car perfect on a winding road, makes it slightly darty when you're adjusting the radio on the freeway. The Miata likes drivers who pay attention to what they are doing. While the steering ratio feels faster at center than it does toward extremes, the overall ratio is perfectly suited to the car's wheelbase and weight. On a winding road, the Miata truly feels like an extension of the driver's nervous system.

Willing though the engine is, the 1.6-liter dohc four is no torque monster. Drop below 3000 rpm on an uphill hairpin, and you'll think the throttle pedal is connected to the vacuum system. Spin the engine from 4000 up toward its 7000-rpm redline, however, and it moves the car out in nice surges of acceleration, emitting one of the best exhaust notes this side of Italy.

Top up, the car is snug and reasonably quiet for a convertible; top down, only the tallest drivers were aware of a little wind around the upper dreadlocks. The seat is comfortable and supportive, there's a dead pedal for the left foot, heel-

Mazda MX-5 Miata

PRICE	
Base price	**$13,800**
Price as tested	**$15,740**
SPECIFICATIONS	
Curb weight	**2205 lb**
Wheelbase	89.2 in.
Track, f/r	55.5 in./56.2 in.
Length	155.4 in.
Width	65.9 in.
Height	48.2 in.
ENGINE & DRIVETRAIN	
Engine	dohc 4-valve/cyl **inline-4**
Bore x stroke	78.0 x 83.6 mm
Displacement	**1598 cc**
Horsepower (SAE)	**116 bhp @ 6500 rpm**
Torque	**100 lb-ft @ 5500 rpm**
Transmission	**5-speed manual**
CHASSIS & BODY	
Layout	**front engine/rear drive**
Brake system, f/r	**vented discs/discs**
Wheels	cast alloy, **14 x 5½JJ**
Tires	**P185/60HR-14**
Steering type	**rack & pinion**, power assist
Suspension, f/r: **upper & lower A-arms,** coil springs, tube shocks, anti-roll bar/**upper & lower A-arms,** coil springs, tube shocks, anti-roll bar	
PERFORMANCE	
Acceleration:	
0–60 mph	9.5 sec
0–1320 ft (¼ mi)	17.0 sec @ 81.5 mph
Braking:	
Minimum stopping distance	
From 60 mph	154 ft
From 80 mph	270 ft
Handling:	
Lateral acceleration	0.83g
Slalom (700-ft)	62.4 mph
Fuel economy:	
Normal driving	25.5 mpg

The Miata is the lightest, nimblest, least complicated, least expensive car in the test. A weakness is the engine's shortness of breath at low rpm.

and-toe shifting is possible (with good ankle mobility), and the trunk holds a small suitcase in close company with a small spare.

A majority of the group said the Miata would be their first choice if they had to pick one car for the entire trip. A Feature Editor said, "It has everything I want from a sports car—it's fun, light, quick, agile and it's the cheapest. What more could you ask?"

Mazda RX-7 Convertible

STEP FROM THE Miata into the RX-7, and you've backed one notch away from the sports-car ideal and advanced a corresponding notch in comfort and ride. You'll still feel the road surface in the non-turbo RX-7, but there is considerably more body roll, particularly in fast, sweeping corners. Steering is also a bit slower, so in rapid switchbacks the elbows get more of a workout than they do in the twitchy-quick Miata. Grip is still tenacious, but the entire car has a heavier, softer and more civilized feel. The chassis also feels more flexible than the other cars in the group, particularly in rough, tight corners.

Several drivers thought the RX-7 was the least distinctive-looking car of the group—clean and handsome, but unlikely to be photo-

At left, two cars with very different personalities from the same maker. Above, we bow our heads and extend our hands in a moment of prayer, assuring precisely clipped apexes and a safe journey.

■ **Low steering-wheel position and aggressive seat bolstering defy attempts at an easy entrance to the RX-7's snug interior. But once ensconced, the driver enjoys a position that's hard to fault.**

graphed 20 years from now for its trend-setting style. It seems 944-derived, but without that car's sculpted edges, and it is certainly more bland in appearance than the original RX-7.

Drive the RX-7, though, and you'll find plenty of character, most of which comes from under the hood. The 13B rotary has a wonderful razor-sharp edge to its throttle response and power delivery, accompanied by a nice canvas-ripping exhaust note. Like the Miata, it prefers to be kept well above 3000 rpm when real kick is called for, but working the gear lever through

Mazda RX-7 Convertible

PRICE	
Base price	**$26,530**
Price as tested	**$27,389**
SPECIFICATIONS	
Curb weight	**3030 lb**
Wheelbase	95.7 in.
Track, f/r	57.1 in./56.7 in.
Length	169.9 in.
Width	66.5 in.
Height	49.8 in.
ENGINE & DRIVETRAIN	
Engine	2-rotor **Wankel**
Bore width x chamber axis	80.0 x 240.0 mm
Displacement	**1308 cc**
Horsepower (SAE)	**160 bhp @ 7000 rpm**
Torque	**140 lb-ft @ 4000 rpm**
Transmission	**5-speed manual**
CHASSIS & BODY	
Layout	**front engine/rear drive**
Brake system, f/r	**vented discs/vented discs**
Wheels	cast alloy, **15 x 6½JJ**
Tires	**205/60VR-15**
Steering type	**rack & pinion**, power assist
Suspension, f/r: **MacPherson struts,** lower A-arms, coil springs, tube shocks, anti-roll bar/**trailing arms** with camber-control arms, semi-trailing links, floating hubs, coil springs, tube shocks, anti-roll bar	
PERFORMANCE	
Acceleration:	
0–60 mph	9.1 sec
0–1320 ft (¼ mi)	17.1 sec @ 82.5 mph
Braking:	
Minimum stopping distance	
From 60 mph	145 ft
From 80 mph	245 ft
Handling:	
Lateral acceleration	0.82g
Slalom (700-ft)	63.3 mph
Fuel economy:	
Normal driving	19.0 mpg

its gates on an upstream mountain road is an aural and physical delight. It spins willingly and smoothly to its 8000-rpm redline.

A low, nonadjustable steering wheel and heavily bolstered seats make the RX-7 a bit difficult to enter, but once you are in, the seats are snug and supportive. The interior is a mixed bag of good and bad, the bad being a silly non-detent turn-signal flipper that had every driver going down the road with the directionals blinking for miles, driving everyone else nuts. Also, the orange-on-black instrument markings are harder to read than the black-and-white combinations to which most others have, thankfully, returned. The good part is the driving position; wheel, pedals and shift lever are right where they should be.

Well-rounded might be the most apt description of the RX-7. It has the power and handling to keep up with the best in its class, and enough road comfort not to fatigue the driver on an all-day trip.

Nissan 300ZX

WITH THIS CAR we go up yet another step in mile-eating comfort and quietness. Though the ZX feels weightier and more solid than either of the Mazdas (partly because it's a coupe), there's little compromise in the working of its highly developed suspension. It feels, as one driver put it, "like a big, stable, wide platform with ample rubber at each corner." Of all the cars in our test, it may be the most reassuringly stuck to the pavement and the easiest to drive fast.

At times, it's almost like driving a Group C Le Mans car, with a slightly more civilized set of springs and suspension bushings. Steering is accurate, body roll minimal and balance nearly

The 300ZX was one of the most reassuring cars to drive quickly down an unfamiliar stretch of road. Cornering is flat, grip is impressive and V-6 power is delivered without a hitch.

ideal. Slight understeer sets in first, the rear end stays planted and the car corners hard with supple ease. One driver climbed out and said, "You'd have to be an idiot to put this thing off the road." That put everyone on guard, and no one strayed from the pavement during the trip.

The Nissan had the only 6-cylinder engine in the group, smooth revving but with a little more rumble than the others. Those who had recently driven *Road & Track's* long-term 300ZX Turbo, however, found themselves missing the extra horsepower, and there are times when you feel that the naturally aspirated ZX is carrying too much weight for its engine. It's lively enough under most conditions, but you'll want to downshift to 4th for a quick highway pass. Shift action is second only to the Miata for its precision, and the brake pedal is impressively firm and easy to modulate. Steering is pleasantly linear, adding to the car's great sense of stability, but it is less communicative than the Miata or the Porsche, for instance. One driver noted, "It doesn't talk back to you, but it doesn't talk to you, either."

The ZX gives you a nice molded-in-the-cocoon driving position, with a good low-dash view of the road, though rear outward vision is the worst of the group. Pod-mounted controls are easy enough to use, once you figure out where they are, and the white-on-black instruments are simple and easy to read.

Styling, of course, is always subjective, but nearly everyone thought the ZX was handsome, its rounded, flowing lines suggesting the actual mechanical essence of the car—solid, strong, low, unflappable and highly refined.

Porsche 944

CLIMB INTO THE 944, slam the door with a solid "chunk," accelerate up the road into the curves and two words immediately come to

Nissan 300ZX

PRICE	
Base price	**$28,600**
Price as tested	**$30,500**
SPECIFICATIONS	
Curb weight	**3220 lb**
Wheelbase	96.5 in.
Track, f/r	58.9 in./60.4 in.
Length	169.5 in.
Width	70.5 in.
Height	49.4 in.
ENGINE & DRIVETRAIN	
Engine	dohc 4-valve/cyl **V-6**
Bore x stroke	87.0 x 83.0 mm
Displacement	**2960 cc**
Horsepower (SAE)	**222 bhp @ 6400 rpm**
Torque	**198 lb-ft @ 4800 rpm**
Transmission	**5-speed manual**
CHASSIS & BODY	
Layout	**front engine/rear drive**
Brake system, f/r	**vented discs/vented discs**, ABS
Wheels	cast alloy, **16 x 7½**
Tires	**225/50VR-16**
Steering type	**rack & pinion**, power assist
Suspension, f/r	**angled upper arms, lower A-arms,** "3rd links," coil springs, tube shocks, anti-roll bar/**multimember** (upper lateral links, mid-lateral links, lower angled A-arms), coil springs, tube shocks, anti-roll bar
PERFORMANCE	
Acceleration:	
0–60 mph	7.1 sec
0–1320 ft (¼ mi)	15.5 sec @ 90.5 mph
Braking:	
Minimum stopping distance	
From 60 mph	134 ft
From 80 mph	233 ft
Handling:	
Lateral acceleration	0.88g
Slalom (700-ft)	65.6 mph
Fuel economy:	
Normal driving	19.0 mpg

Porsche 944

PRICE	
Base price	**$48,600**
Price as tested	**$49,215**
SPECIFICATIONS	
Curb weight	**est 3095 lb**
Wheelbase	94.5 in.
Track, f/r	58.2 in./57.1 in.
Length	168.9 in.
Width	68.3 in.
Height	50.2 in.
ENGINE & DRIVETRAIN	
Engine	dohc 4-valve/cyl **inline-4**
Bore x stroke	104.0 x 88.0 mm
Displacement	**2990 cc**
Horsepower (SAE)	**208 bhp @ 5800 rpm**
Torque	**207 lb-ft @ 4100 rpm**
Transmission	**5-speed manual**
CHASSIS & BODY	
Layout	**front engine/rear drive**
Brake system, f/r	**vented discs/vented discs**, ABS
Wheels	cast alloy; **16 x 7 f, 16 x 8 r**
Tires	**205/55ZR-16 f, 225/50ZR-16 r**
Steering type	**rack & pinion**, power assist
Suspension, f/r	**MacPherson struts,** lower A-arms, coil springs, tube shocks, anti-roll bar/**semi-trailing arms,** torsion bars, tube shocks, anti-roll bar
PERFORMANCE	
Acceleration:	
0–60 mph	7.1 sec
0–1320 ft (¼ mi)	15.3 sec @ 91.5 mph
Braking:	
Minimum stopping distance	
From 60 mph	134 ft
From 80 mph	238 ft
Handling:	
Lateral acceleration	0.83g
Slalom (700-ft)	62.7 mph
Fuel economy:	
Normal driving	23.0 mpg

mind: "substance" and "horsepower." Everything about the Porsche suggests ingotlike solidness. The steering wheel has a thick, meaty feel; the shift lever takes a healthy tug to go through its precise gates; and the traditional, communicative kickback through the steering tells you there are large, well-connected tires out at the ends of the tie rods.

Exit a corner with the throttle on and the big counterbalanced 3.0-liter inline-4 responds with locomotivelike torque and pushes you solidly back into the seat. This combination of sophisticated brawn and punch from under the hood makes the 944 marvelously entertaining and satisfying to drive hard on a good sports-car road. It is perhaps the only car in the group

■ The Porsche 944's stiff structure, Herculean 3.0-liter inline-4 and quick reflexes are its greatest assets. While handling is superb and ultimate grip excellent, the tires protest audibly under moderate side loading.

you drive as much with the throttle pedal as with the steering wheel.

Pushed hard, the Porsche's tires start to complain a little earlier than those on other cars in our group, and better tires might make the 944 more reassuring to drive. It has a taut, "rubberized" feel to the suspension, strong brakes and excellent balance, but it also has a sense of mass (weighing in at close to 3100 lb.) that communicates the subtle notion that you wouldn't want this baby to get away from you. It may, in fact, be the Porsche's very heft and solidness that keep it from being quite as much fun to drive as the Miata. The Miata's lightness makes the transition from full adhesion to drift seem less intimidating.

In body structure, the 944 is the least flexible of the convertibles and has the least cowl shake, thanks to the second floor panel Porsche welded in when the top was removed, giving the chassis a sandwichlike cross section to the floor. The steering wheel, though nonadjustable, is nicely placed and controls are all within easy reach (except for the mirror controls that no one can ever find). Top up, rear corner views are a bit restricted for changing lanes, and the car is fairly noisy—though the noises are all good ones, such as intake moan and engine growl.

Overall, the 944 is the car we voted "most likely to be around 20 years from now." It exudes well-constructed quality, and you certainly pay for it. At a sobering $49,215, it is the most expensive car in the group.

Toyota MR2

THE OLD SQUARISH, spritely MR2 has given way to a slightly more serious interpretation of the mid-engine 2-seater theme, and while some lament the loss of the early car's sheer cheekiness, others applaud the new version for looking and feeling a little more like a "real" car. More to the point, the body has been restyled in a frankly Ferrari-like vein, particularly forward of the roof pillars, and it has a whoopy mechanical spirit that also suggests a 2.2-liter interpretation of the Testarossa theme.

Being smaller and lighter than any Maranello product, however, it is considerably more nimble and arguably a lot more fun to drive—at least on this side of 80 or 85 mph. The dohc inline-4 revs to its 6300-rpm redline with the quick willingness of a Formula Atlantic engine (slightly detuned), sounding pleasingly raspy, if a bit

Top-down touring, Alfa Romeo style

For a car introduced when Jack Brabham was still winning Formula 1 driving championships, the Alfa Romeo Spider still has a lot to offer the connoisseur of back roads. "The best trundler of the bunch . . . you don't have to push it to have fun," we said of it in "Back to the Mother Lode" (August 1989), where we pitted old and new versions of the Spider, Porsche 911 and Chevrolet Corvette against one another.

For our current test, the heart-and-soul Alfa would be a perfect complement to our bevy of computer-designed, engineered-to-near-perfection sports cars. Would the veteran of 25 years be able to stay in stride with the new blood from Japan and West Germany? We aimed to find out, but the Spider we had arranged to be waiting for us at a San Francisco airport parking lot was nowhere to be seen.

Not to worry. A couple of days after the test we took delivery of a silver Spider Veloce ($23,273) at our Newport

The MR2's 2.2-liter engine is a mixed bag: It offers surprising torque at very low rpm and a great snarl as it ascends the rev scale, but its boomy, buzzy nature on the overrun is obtrusive.

boomy and buzzy on the overrun. Shift and clutch action are light, as is steering effort, giving a refined, almost delicate feel to the controls.

Ride is the most supple and bump-absorbing of any car in the group, but the MR2 corners with no untoward body sway, quick and flat as a road-going manta ray. The suspension gives the tires a gluelike grip on the blacktop, and the

Beach offices for driving evaluation. The Spider greets 1991 with a Pininfarina restyle, receiving integrated taillights, a subtly shaped sheet-metal lip on the trunklid (replacing the rather heavy-handed black rubber spoiler on last year's car) and smoother, color-keyed bumpers. We all agreed the new additions make the car look smoother of line and less fussy, but some of the aggressive look of last year's car is lost.

Swing open the petite doors (unlatched with handles whose style Mazda liked well enough to use on the Miata), sit down on seats of creamy smooth Italian leather and it's 1966 again. Well, almost. To satisfy government safety regulations, the Spider's steering wheel now incorporates an airbag, and knee restraint bolsters jut out awkwardly from the underside of the dash.

The shift lever, with its longish throws and honest mechanical feel, still sprouts from the console, rather than up from the floor. And the dash is still fitted with those round, double-vane air vents that Italian manufacturers prize. It's a charming, though a bit cramped, place to watch roadside scenery unfurl.

We reckon the Alfa would be breaking a sweat trying to keep pace with the rest on the roads near Bodega Bay. The aging 2.0-liter twincam engine puts out a modest 120 bhp and is asked to push a car weighing nearly 300 lb. more than the Miata. Torque, 117 lb.-ft. produced at a low 2700 rpm, helps the Spider move off the line smartly, but increasing resonances past 4000 rpm or so discourage twisting the tach needle to its upper reaches too often. Gun it for all she's worth, and 60 mph comes up in 10.8 seconds.

The chassis feels somewhat willowy, doing nothing to enhance the steering's lack of crispness at low speeds. As velocity increases, the chassis is increasingly willing to be your partner in fun; favorable 54/46 front/rear weight distribution allows the modest limits of the chassis and tires to be exploited to their fullest.

Put yourself in a relaxed state of mind and put the top down, and you can experience what the Alfa does best: the pleasurable act of unhurried motoring.—*Douglas Kott*

PHOTOS BY RON PERRY

Forget the barrel of monkeys—what could be more fun than five honest-to-God sports cars, challenging roads, spectacular scenery . . . and an oyster bar?

Toyota MR2

PRICE	
Base price	**$14,898**
Price as tested	**$16,723**
SPECIFICATIONS	
Curb weight	**2670 lb**
Wheelbase	94.5 in.
Track, f/r	57.9 in./57.1 in.
Length	164.2 in.
Width	66.9 in.
Height	48.8 in.
ENGINE & DRIVETRAIN	
Engine	dohc 4-valve/cyl **inline-4**
Bore x stroke	87.0 x 91.0 mm
Displacement	**2164 cc**
Horsepower (SAE)	**130 bhp @ 5400 rpm**
Torque	**140 lb-ft @ 4400 rpm**
Transmission	**5-speed manual**
CHASSIS & BODY	
Layout	**mid engine/rear drive**
Brake system, f/r	**vented discs/vented discs**
Wheels	cast alloy; **14 x 6 f, 14 x 7 r**
Tires	**195/60HR-14 f, 205/60HR-14 r**
Steering type	**rack & pinion**
Suspension, f/r: **MacPherson struts,** lower A-arms, coil springs, tube shocks, anti-roll bar/**Chapman struts,** lower lateral links, toe links, trailing links, coil springs, tube shocks, anti-roll bar	
PERFORMANCE	
Acceleration:	
0–60 mph	8.5 sec
0–1320 ft (¼ mi)	16.3 sec @ 84.0 mph
Braking:	
Minimum stopping distance	
From 60 mph	145 ft
From 80 mph	251 ft
Handling:	
Lateral acceleration	est 0.86g
Slalom (700-ft)	63.6 mph
Fuel economy:	
Normal driving	24.0 mpg

car has to be pushed quite hard before mild understeer sets in. A quick lift brings the rear end out a foot or two and then it tucks back in. It's a delightful car to drive on a tight, winding road, yet remarkably comfortable on a nine-hour Interstate cruise, as we discovered. Most of us agreed that for just plain fun of driving, it comes a very close second to the Miata, but makes a more comfortable long-distance partner.

Being mid-engined, the MR2 leaks some engine boominess through the firewall behind your back, particularly in 4th gear, and luggage space is divided between a reasonably deep well for one large (or two small) suitcases at the rear and a small, soft bag at the front. Other than those minor cautions—and the tail treatment, which nobody liked—the MR2 is hard to fault. It's well-finished, has a cleanly designed, user-friendly cockpit and handles like a civilized GTP car. At $16,723, it's one of the bargains in this group.

In good company

WHILE THIS DRIVE was intended as a survey of the naturally aspirated sports cars on the market rather than a direct comparison test (the cars being so far apart in price), comparisons and conclusions are inevitable when you climb in and out of five different cars for four days.

And one conclusion that might be drawn is that you can't fool Mother Nature. If it's sport you are looking for, a light car of small dimensions—suspension and engine technology being equal—is simply more fun and more sporting to drive. If mass is the enemy of performance, however, it has never been the enemy of comfort or quietness. The more heavily built cars give you more of both.

Another inescapable observation is that all five of these cars are remarkably good machines for a hard drive through sports-car country. A similar drive with a batch of sports cars from 20 years ago would have produced varied incidents of wheelspin, axle tramp, brake fade, swing axle hop, tire chunking and engine overheating, if not actual mechanical failure. This current group of cars got through the week without a single mechanical glitch and without any aberrations in handling or braking. They all handle better than most outright racing cars used to, while being a lot more comfortable and reliable.

In the real world of the twisting public highway as we found it in Northern California, no one car was able to pull away and leave the others behind; some were just slightly faster under different conditions. The equality of performance was remarkable. We seem finally to have reached a point where sports cars actually perform as people have always wished they would, and any gathering of the breed automatically puts you in good company.

Mazda RX-7

The Mazda RX-7 competes in a class of one. Sure, there are other sports cars around, almost all available in either coupe or convertible form, but only Toyo Kogyo's speedster has a Wankel engine. It, and its successors when they appear, will likely maintain that distinction.

Most experts and enthusiasts hailed the advent of the rotary powerplant when it was unveiled many years ago, applauding its uncanny smoothness, high-rpm capability and high output for a given displacement. But the development process was a long, drawn-out one, and concerns over exhaust cleanliness, apex-seal reliability and excessive thirst nearly doomed the Wankel.

Mazda, however, stuck to its guns, and kept the rotary in production, though it is now available to U.S. customers only in the RX-7.

That may be the best showcase possible for the engine. As the Wankel has been steadily refined to be cleaner, more powerful and less wasteful of fuel, the RX-7 has been honed to a high level of competence.

Three of the four RX-7 models—GTU, GXL and Convertible—are powered by a normally aspirated twin-rotor Wankel, massaged (in 1990) to put out a healthy 160 bhp. For so powerful an engine, this rotary is quite docile around town. Perhaps a bit too docile at low speeds, making use of the crisp-shifting 5-speed manual gearbox a (pleasant) necessity. The alternate transmission is a 4-speed automatic; this is a suitable choice for the nonturbo RX-7, one good enough to make manual-shift advocates take a second look.

But it's the Turbo II that is the real powerhouse. Two hundred horses are as near as your right foot in this one, along with 196 lb-ft of torque. And the pressure the Turbo puts on your back is as real as its seems: 0–60 mph runs can be made in a 6.4 sec, with another 80 mph or so ahead.

The chassis and brakes are up to demands made on them by either engine. A power-assisted steering rack works with the front MacPherson struts to provide first-class steering response, while the Dynamic Tracking Rear Suspension's floating hubs get the power down. No surprises await; this is a sure-footed sports car. The Turbo is even better, thanks to its stiffer anti-roll bars and wider tires. All RX-7s have 4-wheel disc brakes; Turbos get ABS.

All this goodness is housed in an attractive body that received its last facelift in 1989. The interior is equally good, having all the necessary sports car features in a well-designed package. About all that need be added for maximum luxury are the optional leather seats.

SPECIFICATIONS

Base price, base model	**$19,335**
Country of origin	**Japan**
Body/seats	**3D*, conv/2**
Layout	**F/R**
Wheelbase	**95.7 in.**
Track, f/r	**57.1 in./56.7 in.**
Length	**169.9 in.**
Width	**66.5 in.**
Height	**49.8 in.**
Luggage capacity	**6.5 cu ft**
Curb weight	**2985 lb**
Fuel capacity	**18.5 U.S. gal.**
Fuel economy (EPA):	
City	**16 mpg**
Highway	**24 mpg**
Engine	**160-bhp 2-rotor Wankel**
Optional	**200-bhp turbo 2-rotor Wankel***
Bore width x chamber stroke	
Major axis, mm	**80.0 x 240.0**
Displacement	**1308 cc**
Compression ratio	**9.0:1**
Horsepower, SAE net	**200 bhp @ 6500 rpm**
Torque	**196 lb-ft @ 3500 rpm**
Transmission	**5M*, 4A**
Suspension, f/r	**ind/ind**
Brakes, f/r	**disc/disc**
Tires	**205/55VR-16**
Steering type	**rack & pinion (p)**
Turning circle	**32.2 ft**
Warranty, years/miles:	
Bumper-to-bumper	**3/50,000**
Powertrain	**3/50,000**
Rust-through	**5/unlimited**

*indicates model described in specifications.

Mazda MX-5 Miata

For those who missed out on Mazda's perky MX-5 Miata during its first year of production, there's good news: It's entering a second year of production with a change that makes it even more desirable. Mazda has made ABS an option on its dandy little 2-seater. With that single addition, the Miata is just about perfect.

And for those who resisted the charms of the tiny 2-seater because their driving program includes too much stop-and-go driving for comfort, Mazda has introduced a Miata with an automatic transmission.

Purists, this is your moment to recoil in horror. The Miata, a throwback to the era of the British Sports Car, with a *slushbox*? Actually, the idea makes sense. Replacing the slick 5-speed manual transmission with a 4-speed self-shifter does not deprive the Miata of its basic virtues; it just makes things easier in traffic.

Unchanged is the smooth little body Mazda's designers penned for the Miata. No controversy surrounds the MX-5's looks; the worst anyone we know has said of it is that it is "cute." Others offer more fulsome praise. And the public has reacted by snapping up Miatas as fast as they come off the transport ships.

Same goes for the interior. There is room for two in nicely shaped seats, a handsome instrument panel provides all the information a driver could want in a clear fashion, and there's an airbag in the steering wheel hub, obviating the need for any oddball seatbelt systems.

Also carried over are the peppy little 116-bhp twincam engine and sophisticated chassis pieces of last year, each of which does its best to give Miata drivers that you-and-the-car-are-one feeling.

Popular options continue. Just about everyone ordered air conditioning last year, and many went for the handsome spoked alloy wheels that come with power-assisted steering and a leather-wrapped steering wheel in the A Package. Some went for the B Package, which contains all of the A Package plus electric window lifts, cruise control and head-rest-mounted stereo speakers, and a few called for the optional hardtop. Given the ease of operation and weather-tight nature of the standard soft top, that's not surprising.

Okay, the Miata is great fun to drive, reasonably quick, good to look at, and well built, and gets very respectable fuel mileage. The top boot may flap a bit at speed, but there is just enough trunk space for two for a wonderful week of touring.

The Miata's strength has been its appeal to the carefree types. Now it can charm the shiftless, too. And don't forget to order the ABS.

SPECIFICATIONS

Base price, base model	**$13,800**
Country of origin	**Japan**
Body/seats	**conv/2**
Layout	**F/R**
Wheelbase	**89.2 in.**
Track, f/r	**55.5 in./58.2 in.**
Length	**155.4 in.**
Width	**85.9 in.**
Height	**48.2 in.**
Luggage capacity	**6.0 cu ft**
Curb weight	**2205 lb**
Fuel capacity	**11.9 U.S. gal.**
Fuel economy (EPA):	
City	**25 mpg**
Highway	**30 mpg**
Engine	**116-bhp dohc 16V inline-4**
Bore x stroke	**78.0 x 83.6 mm**
Displacement	**1597 cc**
Compression ratio	**9.4:1**
Horsepower, SAE net	**116 bhp @ 6500 rpm**
Torque	**100 lb-ft @ 5500 rpm**
Transmission	**5M*, 4A**
Suspension, f/r	**ind/ind**
Brakes, f/r	**disc/disc**
Tires	**P185/60HR-14**
Steering type	**rack & pinion**
Turning circle	**30.0 ft**
Warranty, years/miles:	
Bumper-to-bumper	**3/50,000**
Powertrain	**3/50,000**
Rust-through	**5/unlimited**

*indicates model described in specifications.

PHOTOS BY DEAN SIRACUSA

MAZDA RX-7 INFINI IV

A preview of the new RX-7, now playing in the current platform

"HEY, YOU'RE SITTING on the wrong side!" It's a common reaction from Californians when they see this right-hand-drive Mazda RX-7 Infini IV at a stoplight. It's more than a little odd to be a handshake's distance from the driver in an adjacent lane; it's unsettling to grapple with lane position when your body is in the "passenger" seat—one of us likened the feel to riding a motorcycle with a sidecar for the first time, always worried about what that dang appendage is going to hit. And we all taught our left arms new tricks in muscle memory trying to shift with something resembling finesse.

But we digress. Mazda's intent in lending us this Japanese-market-only Infini (pronounced on′-fin-ee) goes far beyond novelty. It represents the ultimate production RX-7, and it is a sample of what the next-generation RX-7 (due out in 1992) will be like in handling and ride.

Details on the new RX-7 remain sketchy, but we know it'll be slightly shorter, wider and lighter than the present car. Power will come from the evergreen 13B 2-rotor Wankel, in sequential twin-turbocharged form only, with "substantially more" power than the present Turbo's 200 bhp. Within the framework of the current platform, the Infini paves the way for a leaner, more elemental driver's RX-7, a return to the basic mechanical directness of the RX-7s that first reached U.S. shores in 1979.

Following the race driver's adage, "To go faster, add lightness," Mazda's engineers, headed by Takaharu Kobayakawa, have created an RX-7 that weighs about 265 lb. less than a U.S.-spec RX-7 Turbo. First to go were the stock seats, replaced by lighter, deep buckets, with flared upper seatbacks. Their snug-fitting nature makes the already cozy RX-7 cockpit seem more intimate, as do molded rectangular pads on the driver's-side door panel and center console, to brace one's legs while cornering. The passenger is not

■ **Turbocharged 13B rotary makes 215 bhp JIS, up slightly from stock through a lower-restriction exhaust system, but the Infini's top speed is limited to only 112 mph. The interior looks suited for twice that speed, with high-bolstered racing-type seats and a 3-spoke Momo wheel whose stitching and leather cover feel just so.**

ignored when it comes to g-loads either; a "dead pedal" bar running the width of the footwell is provided.

A Momo 3-spoke steering wheel frames the huge center-mounted tach, which is flanked by a 180-km/h speedometer and a cluster of gauges for fuel level, coolant temperature, oil pressure and boost level. Other touches include an emergency flare that clips into a holder under the glovebox, and a perforated suedelike material that covers the seats and inset portions of the door panels.

PAMPERED THE INFINI driver is not, because such amenities as air conditioning, sound system, central locking, a sunroof, cruise control, rear-window wiper and the luggage-area partition are tossed out for the greater good of lighter weight and consequent more nimble handling. Also toward this end, the Infini gets a suspension makeover, with its front MacPherson-strut/rear modified-trailing-arm setup firmed and toned with stiffer springs, thicker anti-roll bars and significantly harder bushings. Specially tuned Pirelli P Zero tires, in the standard Turbo size of 205/55ZR-16, are mounted on forged BBS wheels, each responsible for a slight weight savings over the stock wheels.

Under the hood, there's that same little beer keg of an engine, the 1308-cc twin-rotor Wankel, set well back in its compartment to help achieve 49/51 weight distribution. Buried beneath the air-to-air intercooler, induction plumbing and other clutter is a twin-scroll turbo, a device whose internal ducting lets each rotor's exhaust pulse have a separate, unfettered shot at the impeller to lessen turbo lag and make for more tractable turbo power. On the exhaust side, a lower-restriction system helps to bump up power to 215 bhp JIS, and torque by 10 lb.-ft. to 206. But the eye-grabber is the bright red strut-tower brace that reiterates the Infini's dedication to precise suspension geometry.

This engine and chassis work together with an ease that's surely not accidental. Where the stock RX-7 Turbo takes a noticeable understeering set when pushed to its limits, the Infini seems as neutral as Switzerland; we suspect the firmer bushings in the rear suspension's floating-hub arrangement hold to a bare minimum its understeer-producing toe-in during lateral acceleration.

The Pirelli P Zeros, with a tread compound sticky enough to fling gravel into the wheel wells at parking-lot speeds, have enormous grip (we saw 0.91g on the skidpad) and approach their limits in a predictable manner. The tail of the U.S.-spec RX-7 Turbo can be wrestled out by heavy throttle application in 2nd or 3rd gear; with the Infini it's more of a waltz than a wrestling match—each tire contributes more equally to the task of scurrying down the road, and all that's required to change the car's aim (provided there's enough side loading) is a caress of the throttle.

Though there is no sound system per se, the Infini generates lots of interesting noises anyway. With the rotary's eagerness to rev, the tach's warning buzzer is heard often. And there's the shrill song of the turbo—more apparent because of skimpier body sound-deadening—that rises in pitch faster than the whine of the engine. The starch-stiff sidewalls of the Pirellis drum and thump over poor road surfaces, turning the interior into a personal resonation chamber.

But such minor cacophony is forgiven when the rewards are considered. The rack-and-pinion steering uses an engine-speed-sensitive assist that feels more natural than the Turbo's, with more feedback at lower speeds. The brake system, while retaining vented discs all around (10.9 in. front, 10.8 in. rear), makes do without ABS but receives Infini-specific front pads and other minor alterations that reduce brake pedal stroke by 3/8 in. and give the brake pedal appreciable (and appreciated) firmness. Stopping distances were somewhat shorter than those of the last RX-7 Turbo we tested (August 1988): 8 ft. less from 60 mph, and 14 ft. less from 80.

Surprisingly, what with less weight and slightly more power, acceleration times were not as quick as expected—7.0 seconds to 60 mph and 14.9 sec. through the standing quarter mile. A bit of wheel tramp off the line, a possible consequence of the sticky-compound Pirellis, could have slowed the Infini down a few tenths.

But that shouldn't be a bother. Think of the Infini as a rare treat for the Japanese market and a limited-production test bed pointing the way to a leaner, meaner production RX-7 that will roll into U.S. showrooms in 1992. Through the years, we've seen sports cars grow larger, softer and heavier, which makes Mazda's reversal of the trend even more exciting. Let's hope that this Japanese-market Infini's chassis is truly representative of the new generation's, and that Mazda's stylists have created a shape worthy of the mechanicals.

Test Notes . . .

- Both slalom and skidpad testing showcased the Infini's firm springing and resistance to body roll. Its steering is quick—a nice feature because the Infini's tail can step out abruptly.
- Acceleration and slalom times were probably hindered a mite by our unfamiliarity with the Infini's right-hand driving position. With practice, both would surely improve.

MAZDA RX-7 INFINI IV

0–60 mph	7.0 sec
0–¼ mi	14.9 sec
Top speed	est 112 mph
Skidpad	0.91g
Slalom	60.7 mph
Brake rating	very good

PRICE

List price, Japan .... **est $21,000** Price as tested .... **est $21,000**

Price as tested includes std equip. (BBS wheels, sport seats, Momo steering wheel, elect. adj windows & mirrors).

ENGINE

Type	turbo, **2-rotor Wankel**
Displacement	79.8 cu in./1308 cc
Bore width x chamber major axis:	3.15 x 9.45 in./80.0 x 240.0 mm
Compression ratio	9.0:1
Horsepower (JIS):	**215 bhp @ 6500 rpm**
Torque	**206 lb-ft @ 3500 rpm**
Maximum engine speed	7000 rpm
Fuel injection	Mazda elect. port
Fuel	unleaded, 87 pump oct

GENERAL DATA

Curb weight	**2770 lb**
Test weight	2925 lb
Weight dist, f/r, %	49/51
Wheelbase	95.7 in.
Track, f/r	57.1 in./56.7 in.
Length	**169.9 in.**
Width	**66.5 in.**
Height	**49.8 in.**
Trunk space	19.0 cu ft

DRIVETRAIN

Transmission .... **5-sp manual**

Gear	Ratio	Overall ratio	(Rpm) Mph
1st	3.48:1	14.98:1	(7000) 32
2nd	2.02:1	8.66:1	(7000) 56
3rd	1.39:1	5.98:1	(7000) 81
4th	1.00:1	4.30:1	(6950) 112
5th	0.72:1	3.09:1	est (5000) 112

Final drive ratio 4.30:1

Engine rpm @ 60 mph in 5th 2675

CHASSIS & BODY

Layout	**front engine/rear drive**
Body/frame	unit steel
Brakes, f/r	**10.9-in. vented discs/ 10.8-in. vented discs;** vacuum assist
Wheels	forged alloy, **16 x 7J**
Tires	Pirelli P Zero, **205/55ZR-16**
Steering	**rack & pinion,** power assist
Turns, lock to lock	2.7
Suspension, f/r:	**MacPherson struts, lower A-arms,** coil springs, tube shocks, anti-roll bar/**trailing arms, lower lateral links, semi-trailing links,** floating hubs, coil springs, tube shocks, anti-roll bar

FUEL ECONOMY

Normal driving	14.5 mpg
EPA city/highway	na
Fuel capacity	18.5 gal.

INTERIOR NOISE

Idle in neutral	53 dBA
Constant 70 mph	75 dBA

ACCELERATION

Time to speed	Seconds
0–30 mph	2.2
0–60 mph	7.0
0–80 mph	10.8
Time to distance	
0–100 ft	3.1
0–500 ft	8.2
0–1320 ft (¼ mi)	14.9 @ 94.0 mph

BRAKING

Minimum stopping distance	
From 60 mph	141 ft
From 80 mph	242 ft
Control	very good
Pedal effort for 0.5g stop	16 lb
Fade, effort after six 0.5g stops from 60 mph	16 lb
Brake feel	excellent
Overall brake rating	very good

HANDLING

Lateral accel (200-ft skidpad)	0.91g
Balance	mild understeer
Speed thru 700-ft slalom	60.7 mph
Balance	neutral

Subjective ratings consist of excellent, very good, good, average, poor; na means information is not available.

BROOKLANDS BOOKS

ABOUT BROOKLANDS BOOKS

As a company Brooklands Books are dedicated to preserving motoring literature for enthusiasts. They have in print over 700 titles which deal with various aspects of motoring. A guide to each series is given below.

REGULAR BROOKLANDS BOOKS. Collected road tests, model descriptions, driving impressions and other articles on individual models and marques - plus American muscle car compilations - taken from a wide variety of car magazines in Britain, North America and Australia. Filled with informed and reliable comment, specifications, performance data, buying used reports and other details, these books provide a fine opportunity to absorb all relevant information on the car you own, are thinking of buying or restoring, or perhaps simply admire.

GOLD PORTFOLIO SERIES. Again drawing on material from a wide range of motoring magazines, books in this series are devoted to prestige and high-performance cars from Britain, the Continent and North America. They have almost twice as many pages and illustrations as Regular Brooklands Books and are an unparalleled source of interest and knowledge on the world's finest and most sought-after cars.

BROOKLANDS BOOKS "COLLECTIONS". These books are similar to those in the Regular Brooklands Books series, but fewer pages and are lower priced. In cases where both series cover the same marque or model, there is no duplication.

PRACTICAL CLASSICS SERIES. Comprehensive restorers' guides, compiled from the pages of "Practical Classics" magazine and based on "hands-on" experience. All restoration tasks, tools and techniques are covered, together with a large number of instructional photos. Ideal practical aids for both skilled and new restorers, who will find themselves saved much unnecessary time and expense.

ROAD & TRACK SERIES. Collected tests, comparisons, new model introductions, racing and touring articles and other material from the authoritative and internationally popular American magazine "Road & Track".

CAR & DRIVER SERIES. Similar collections from "Car & Driver" magazine, USA. Once again, these books supply a wealth of information on numerous models, in a package easy both to read and to keep.

HOT ROD ON GREAT AMERICAN ENGINES SERIES. Musclecar and high performance material drawn from Hot Rod, Car Craft and other Petersen publications on how to maintain, tune, repair, restore and modify the popular power plants of the musclecar era.

MOTOR and THOROUGHBRED & CLASSIC CAR SERIES. Collected articles from these two highly-regarded British magazine, profiling their subject cars from many angles.

BROOKLANDS TECHNICAL BOOKS. These are all originals or unedited reissues of factory publications, including workshop manuals, handbooks and parts catalogues. Far more detailed than the "condensed" literature widely available, they are essential possessions for owners and restorers and are highly recommended. We also include some official sales brochures within this series.

OWNERS WORKSHOP MANUALS. Brooklands Books have acquired the rights to the Autobooks Workshop Manual series plus some manuals originally published by Chilton and Clymer. Written and illustrated by experts, they cover servicing, overhaul and repair for DIY enthusiasts. Many are now reissued in larger format with an all-new chapter on buying a used model or as handy 'glovebox' editions.

BROOKLANDS GENERAL RESTORATION AND TECHNICAL BOOKS. Practical guides for owners wishing to restore or upgrade their cars. Most of these titles are drawn from the pages of "Hot Rod" magazine and fall into our "Restoration Tips & Techniques" or "Performance Tuning" categories. All contain detailed and expert text, together with hundreds of explanatory photos.

BROOKLANDS REISSUES OF MAGAZINES. Early copies of "Road & Track", "Hot Rod" and "Autosport". Economically priced, these reissues are a boon to collectors who have hitherto found them almost impossible to locate or purchase.

Send for catalogue and full list to:
BROOKLANDS BOOKS LTD, PO Box 146, COBHAM, SURREY KT11 1LG, ENGLAND
Telephone: 0932 865051 Fax: 0932 868803
MOTORBOOKS INTERNATIONAL, OSCEOLA, WISCONSIN 54020, USA
Telephone: 715 294 3345 & 800 826 6600 Fax: 715 294 4448